Volume 1: Materials

Particleboard

A. A. Moslemi

SOUTHERN ILLINOIS UNIVERSITY PRESS
Carbondale and Edwardsville

Feffer & Simons, Inc.
London and Amsterdam

Library of Congress Cataloging in Publication Data

Moslemi, A   A   1935–
    Particleboard.

    Bibliography: v. 1, p.
    CONTENTS: v. 1. Materials.
    1. Particleboard. I. Title.

TS875.M63   674.8   74-2071

ISBN 0-8093-0655-7

Contents

VOLUME 1: MATERIALS

9 Dimensional, Thermal, and Acoustical Properties   137

Thickness Movements 137;   Linear Movements 146;   Dimensional Stabilization 150;   Thermal Properties 155;   Acoustical Properties 157;   Notes 161

10 Moldings and Portland Cement-bonded Products   163

Definitions and Particularities of Moldings 163;   Properties of Moldings 165;
Flow 170;   Portland Cement 174;   Wood-Cement Bonding 178;   Portland
Cement-bonded Products 182;   Notes 186

Contents

Preface

In its relatively short existence, particleboard has gained prominence as a wood product. Research and development to refine particleboard technology has been intense over the last quarter century. Particleboard manufacture is now among the most automated industries. Various processes are still being continuously refined at a rather rapid pace. Many concepts and techniques have become obsolete in a few years. This is expected to continue. It is without doubt that particleboard is here to stay due to its raw material requirements, manufacturing advantages, and product properties.

In spite of much research and development activities, no comprehensive volume has appeared in English in the past on the subject of particleboard. The author began to feel a need for such work to be used as a text as well as a source of reference back in 1965. It was felt then that the amount of knowledge gained since the Second World War was sizable but scattered. Thus, preparation began in that year and work continued until 1970. Much information was examined during the five-year period. Intense writing took place in the academic year 1970–71 during the author's sabbatical leave. The review and revising of the manuscript took place mostly in 1972. Two volumes were prepared with the first dealing with materials and the second with technology. Both a glossary of terminology used and a table of unit conversions have been compiled and appear in the back of each volume. A comprehensive bibliography, which was updated as of March 1973, has also been included to aid the users of these volumes should they wish either to explore certain subjects in greater depth than that included in the text or subjects which have been left out.

All sources cited in short form in the notes at the ends of chapters in each volume and with the illustrative figures and tables scattered throughout the text of each volume are listed in full in the Bibliography.

The author wishes to express his appreciation to a number of specialists for their review of all or parts of the manuscript. Included are Dr. William

Lehmann, Professor Eugene Landt, Mr. Richard D. Putnam, and others.
Special thanks are due Mr. H. John Knapp, Mr. Horst Pagel, and Mr. K.
M. Horner for permitting the author to modify and reuse their previously
published papers on coatings, moldings, and packaging respectively.
Mr. T. M. Maloney also kindly gave his permission as the publisher for
the use of the above three papers. The author is grateful to the many
industrial organizations and scientists whose illustrations and ideas have
been utilized. Ms. Carolyn Weiss and Dr. Dan Irwin of the Southern
Illinois University Cartographic Laboratory deserve special thanks for
spending a number of man-months doing the graphic work.

A. A. Moslemi

Carbondale, Illinois, U.S.A.
May 1973

Volume 1: Materials

Introduction

1–1. Development

The field of wood particleboard is not more than a few decades old. In the United States, unsuccessful efforts were made in the early 1920s to manufacture particleboard.[1] The failure was primarily due to the lack of suitable adhesives: hide glues did not prove expedient from production and economic viewpoints. The desire to utilize profitably the otherwise useless and burdensome residue had inspired the research at that time. Thus, the development of viable particle panels awaited new findings on industrial applications for the thermosetting synthetic resins. New techniques discovered in the 1930s in resin applications paved the way for the industrial production of particleboard in the early 1940s. The first industrial production of particleboard using synthetic resins is believed to have occurred in 1941 in Bremen, Germany,[2] using phenolic binders and spruce particles, although some credit Czechoslovakia with having the first plant some five years earlier. The first flax straw board was commercially produced in 1947 in Belgium.[3]

During the Second World War, the particleboard industry was set back in development due to a resin shortage. War industries consumed substantial quantities of synthetic resins for the production of defense items. Postwar years brought a great desire for reindustrialization and made synthetic resins available for general industrial use. This development stimulated the growth of the wood particleboard industry which has been associated with a great deal of dynamism ever since, both in western Europe and later in the United States.

The European know-how in the field of wood particleboard was received with enthusiasm in the United States in the mid-forties. The first full-scale operation began in the East in 1945 with the establishment of a plant capable of producing 2 million square feet ($\frac{3}{4}$-inch thick basis) annually. Early plants were "captive" and intended to utilize residue from larger operations in order to produce boards for internal consumption. Soon after, the industry grew independent, using not only residues but roundwood as well.

The recognized advantages of particleboard manufacture (see section 1–4) have resulted in a phenomenal rate of growth for this industry in the United States with over sixty plants built in the last fifteen years. Particleboard, over the last twenty years, has emerged from something cynically called "crumble board" to a family of engineered products capable of meeting many requirements. Research performed in this country and in western Europe during this period has been responsible for marked improvements in quality. The industry is continuing to undergo unusually brisk development with the outdating of many ideas in just a few years. This pace of innovation is expected to continue as new markets demand more exacting specifications. Today, there is hardly a furniture plant which does not use substantial quantities of particleboard. In housing and general industrial markets, record quantities are being used year after year.

1–2. Manufacture in Brief

There are two major processes for making particleboard. In the flat-press process, the board is pressed flat-wise; in the extrusion process, it is continuously extruded through a hot die. Both processes have the following basic steps in their manufacture:

1. Reduction of the wood raw material to the desired particle size and shape. This is accomplished by hogging, grinding, hammermilling, or flaking.
2. Drying the particles to a predetermined and uniform moisture content.
3. The oversized and fine particles are separated by screening or some other means of particle segregation. Control is thus exercised over the size and geometry of particles going into the board structure. Often, the fine particles are later deposited on the flat-press boards so that a smooth surface is generated. The coarse particles are redirected into the reduction system for further refinement.

4. Blending (addition and mixing) of calculated amounts of adhesive binder and other additives by spraying or other means. Urea-formaldehyde and phenol-formaldehyde adhesives are currently the major binders used.

5. Forming the blended particles into a "mat" in the flat-press processes. Forming can be controlled so that coarser particles in the furnish are placed in the center thickness of the board while finer particles are deposited closer to the surfaces (graduated layering). Or, the mat may be laid as a randomly deposited single layer mat producing homogeneous boards. In some plants, the mat is deposited such that the surface layers are separately laid.

6. In the hot press or the extrusion die, the mat (flat-press process) or the blended furnish (extrusion process) is consolidated under controlled heat and pressure to a given density.

7. As the hot board emerges from the press, preparatory operations such as cooling, trimming and moisture equalization may be performed.

8. Sanding or planing imposes close thickness tolerances. The generated dust or shavings are either burned or redirected into the process.

9. Further operations are also undertaken in accordance with market demand. These may involve cutting to size, overlaying, routing or filling the surface.

Flat-press boards account for over 95 percent of the total particleboard production. The flat-press process produces a large variety of boards due to the versatility of its engineering design and layout and the large variety of particle shapes and sizes it can accept. The extrusion process normally utilizes hammermilled particles, and its variations of plant design and layout are very limited. The density range is also much wider in the flat-press process than in the extrusion process. Flat-press boards are made with densities as low as 25 pcf to as high as 75 pcf.[4] However, the bulk is currently being produced in a density range from 35 to 50 pcf. Particle sizes accommodated in the flat-press process range from very coarse wood chips to fibers. Particleboard made from wood fibers is commonly called medium-density fiberboard.

The properties of particleboard depend on many factors. The major factors involve the type and size of the particles, techniques of manufacture, kind and amount of resin, particle distribution and orientation, board density, quality of manufacture (i.e., effectiveness of resin spread, forming, etc.), furnish moisture content, and postmanufacturing treatments. In strength, stiffness, and dimensional stability, most flat-press boards are more uniform over the plane of the board compared to extruded boards. The strength of the latter in the extrusion direction is relatively low; thus, the board is always overlaid with veneer to overcome this deficiency before it is used as corestock. Thickness, however, is more stable in the extruded particleboard compared to those made by the flat-press process. The plant layout in the extrusion process is less complex and requires considerably less initial investment. However, due to the inherent deficiencies of this process coupled with accelerated improvements in the flat-press process, the extrusion process is becoming increasingly less favored.

1–3. Production and Markets

The increase in the consumption rate of particleboard has been sharp (fig. 1–1). In the United States, production rose from 111 million square

1–1. *Particleboard consumption ($\frac{3}{4}$-inch basis) in the United States. Consumption for 1975 is based on projected estimates.*

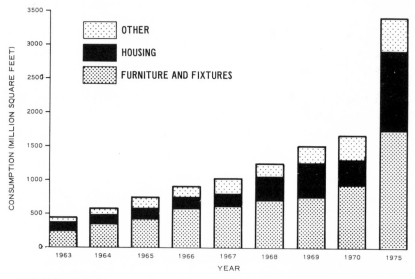

SOURCE: Vajda, P. 1970. The economics of particleboard manufacture revisited or an assessment of the industry in 1970.
Redrawn in modified form by S.I.U. Cartographic Laboratory

feet ($\frac{3}{4}$-inch thick basis) in 1956 to 1.1 billion square feet in 1967 with no indication of slowing down. The estimated production for 1975 is over 3 billion square feet. So far, manufacturing has absorbed most of the production. Well over two-thirds of all particleboard goes into manufacturing, with construction consuming about a quarter.

The use of particleboard in construction will likely increase in the seventies as new techniques and materials are developed. Further refinements in the exterior-type wood particleboard are likely to be favorably received in the field of construction. At the present time, mobile homes consume large quantities of particleboard. In the United States, the need to rectify the housing shortage in the seventies and beyond has been firmly established. In meeting this need, availability of enormous quantities of building materials will be required. Among these, particleboard will probably picture prominently both in interior and exterior applications. The refinement in exterior-type boards is likely to be given further stimulus as the need for an exterior-type panel material at low cost is emphasized. Additionally, increase in the number of new houses built should bring with it increased furniture sales, already a major market for particle materials both in the form of particleboard and moldings.

In summary, with the general growth of the economy in the United States, new markets are likely to be found, particularly in the light of the great production capacity being planned. These markets will probably be captured through development of new products and applications. The particleboard industry is likely to retain its present markets through offering competitive pricing and acceptable quality. Future research and development could, in fact, favorably affect the prices and quality, placing particleboard products at a position where they can capture a yet bigger share of their traditional market. Production will probably reach 4.5 to 5.0 billion square feet (all on $\frac{3}{4}$-inch thick basis) by 1980.

1–4. Assets

During the past few decades, considerable interest has been displayed in western Europe and the United States in profitably manufacturing products from residue otherwise wasted. Waste wood and other lignocellulosic residues have in the past been disposed of by inefficient burning. This is becoming intolerable due to the pollution it generates. Awareness of environmental quality coupled with the need for intensified forest conservation make imperative the intelligent use of residues such as shavings, sawdust, trimmings, scraps, bark, and logging waste. This applies not only to softwoods, but also to low-quality hardwoods since significant portions of the forest area of the United States, as well as that of the world, are a mixture of low-grade hardwoods.

Particleboard has begun to play an important role in the field of residue and low-grade wood utilization. It is being manufactured from residue generated by sawmills, stud mills, and plywood plants as well as from species and grades previously classed as non-commercial. Particleboard is being made not only from residue but also from engineered particles (often from roundwood). The manufacture of particleboard has considerably increased the yield obtained from the harvested trees. The type and quality of residue consumed by the particleboard industry are such that no other economically significant industry has desired it in the past. This is beginning to change, however, since pulp mills are becoming more interested in the use of sawdust and shavings. The effect of an integrated operation has been a marked improvement in the operating economy of the forest industry and a significant step forward in the field of wise resource use. The yield (the ratio of raw materials going into the operation to the salable products emerging) in particleboard manufacture is highly favorable. It is not uncommon to convert over 90 percent of the incoming raw material into finished products. Chemical pulp may yield as little as 45 percent and lumber and plywood yield less than 50 percent.

The advantages of manufacturing particleboard not only lie with the variety of the raw materials which can be used but also in the method of manufacture and the product properties it can offer. The relatively low investment required for the plant (compared to paper and wet-process hardboard) and the degree of automation it can incorporate make particleboard highly competitive. The properties are so engineered as to meet readily many service requirements. For instance, a medium-density particleboard can be designed so that it possesses acceptable dimensional stability, low water absorption characteristics, flat and smooth surfaces, sufficient bending strength, excellent adhesion bonding properties, and good machinability. Particleboard can be made in large panel sizes, with a full range of thicknesses. With the introduction of continuous presses, endless sheets emerge and are subsequently cut into any length desired.

Notes

1. Efforts to make particleboard out of residue in the United States date back to as early as 1901, to those of Henry Watson of Valparaiso, Indiana, who applied for and was issued a United States patent in 1905. However, no commercial venture ever resulted from this early effort. Other patents dealing with particle composites date back to second half of the nineteenth century.

2. Hunt, M. O. 1962. Particleboard industry: Facts and references.

3. Anonymous 1957. Fiberboard and Particleboard.

4. Pounds per cubic foot.

Wood Parameters

2–1. Density

Wood density has significant influence not only on product properties but also on processing. Particleboard made from lower-density species has a greater bending strength, internal bond, modulus of elasticity, and tensile strength, although screw withdrawal resistance, water absorption, and thickness swelling are little affected (figs. 2–1 and 2–2). It may seem illogical that a wood of low density (e.g., aspen or white pine) produces a particleboard with higher strength compared to a wood of higher density (e.g., beech or oak). The reason for this lies in the fact that a given weight of particles from a lightweight wood species occupies a greater volume than the same weight of similar particles from a dense wood. When these volumes of wood are compressed to the dimensions of a board, a higher relative contact occurs for the low-density wood (due to a greater mean compression ratio)[1] resulting in a better adhesive bond between the particles. Greater particle contact promotes better resin efficiency; a critical factor particularly in low-density boards. To be sure, there is a greater amount of adhesive spread per unit surface area on denser particles, but higher relative contact in the lighter particles is the controlling factor for the strength properties of the medium-density particleboard. For high-density particleboard, the adhesive spread per unit surface area of the particles will be the controlling factor of the board strength.

7

2–1. *The relationship between wood density and the strength of the resulting particleboard. The density of particleboard is recorded on the curves.*

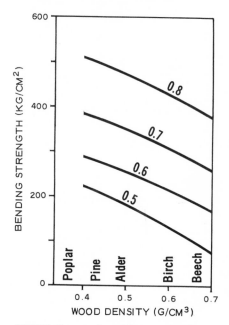

SOURCE: Stegman, G., and Durst, J. 1964. Beech pressboard.
Redrawn in modified form by S.I.U. Cartographic Laboratory

To obtain a given set of strength values with high density wood, the board density must be increased. Figure 2–1 shows the relationship existing between the bending strength of the finished board and the wood density from which the board is made. This figure can be used to estimate how much higher the board density should be when a species of high density is used in the manufacture. For instance, if a particleboard is to be made with 60 percent beech (density 0.68 gm/cm^3) and 40 percent pine (density 0.43 gm/cm^3), the mixture will possess a combined density of 0.4 (0.43) + 0.6 (0.68) = 0.58 gm/cm^3. Figure 2–1 shows that to make a particleboard of equal bending strength to that of a pine particleboard with a density of 0.56 gm/cm^3, the density of the mixed board should be increased to 0.62 gm/cm^3.

The increase in board density when using such species as beech, oak, and hickory implies an increase in expenditure for wood raw material. The increased consumption of wood is compensated for by the increase

2-2. *Relative change in the properties of spruce particleboard with increasing proportion of beech in the particle mixture.*

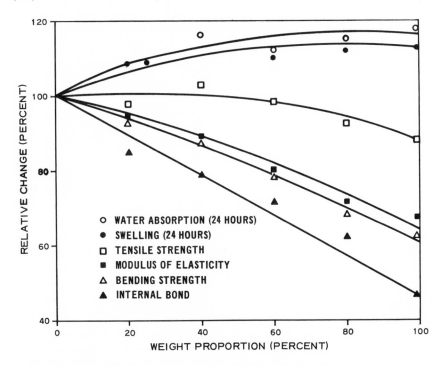

SOURCE: Klauditz, W. 1952. Investigation of the suitability of different types of wood, especially copper beech for the manufacture of particleboard.
Redrawn in modified form by S.I.U. Cartographic Laboratory

in the yield of board per any given cubic foot of original wood, if wood raw material is purchased by volume.[2] As an example, let me consider producing a board which has a bending strength of 3500 psi. Let me further assume that a board density of 0.60 gm/cm[3] is required if a given species of pine is utilized. However, if beech is used, the board density must be increased to 0.70 gm/cm[3] in order to achieve the same strength. The increase in density requires a 16 percent increase in expenditures for wood and binder per unit volume of board produced. The higher volumetric yield in beech particleboard (about 1.15 ft.[3] per ft.[3] of wood) will counter the increase in expenditures compared to pine particleboard (about 0.85 ft.[3] per ft.[3] of wood). A 35 percent increase in yield per cubic foot for beech is gained compared to that of pine. This, coupled with the lower price for hardwoods and an increased usage of pressure refining, points to the feasibility of utilizing hardwood resources.

Mixtures of wood species with various densities behave according to their average density. The mean wood density g_{wm} of the particle mixture can be calculated:[3]

$$g_{wm} = g_{w1}\frac{P_1}{100} + g_{w2}\frac{P_2}{100} + \ldots + g_{wn}\frac{P_n}{100} = \frac{1}{100}\sum_{i=1}^{n} g_{wi}P_i. \qquad (2-1)$$

In equation 2–1, g_{wi} is the oven-dry density of the ith species and P_i is the percent of the ith species in the mix to produce a unit volume of the board. Equation 2–1 can be utilized to calculate the densities of individual species (assuming that the percentages are known) or percentages (if the densities are known). For example, if three different wood species with the ratios P_1, P_2, and P_3 are to be used to make a particleboard of density g_p (oven-dry basis) corresponding to the mean wood density g_{wm}, the densities of the three wood species then are required to be:

$$g_{w1} = 1/P_1(100g_{wm} - g_{w2}P_2 - g_{w3}P_3)$$
$$g_{w2} = 1/P_2(100g_{wm} - g_{w1}P_1 - g_{w3}P_3)$$
$$g_{w3} = 1/P_3(100g_{wm} - g_{w1}P_1 - g_{w2}P_2).$$

These three equations permit an infinite number of solutions. Therefore, it is advisable to think of the density of the two types of wood as being known and solve for the third. The use of these equations, for instance, points out that a mixture of spruce and beech particles with an average density of 38 pcf is just about as suitable for particleboard manufacture as is birch with the same density. Mixing high- and low-density species is not without problems. When air separation is used, for instance, the same sized particles cut from low-density woods will tend to fly farther than particles from denser woods.

The significant use of softwoods in the United States and Europe for particleboard manufacture is largely due to their availability and favorable density. In many instances, however, the use of dense hardwoods either alone or in mixtures with lighter hardwoods or softwoods makes economic sense due to their ready supply and favorable price. This fact has resulted in a rise in the use of hardwoods both in the United States and elsewhere. This increase in usage is very likely to continue for the foreseeable future. The use of hardwoods not only involves low cost and high yield but it also requires lower compression to produce a given board density. Fibrous high quality particleboard from all hardwoods is now being manufactured in large quantities without any difficulty.

2–2. Species

The number of wood species currently utilized in the manufacture of particleboard is numerous throughout the world. In North America, the particleboard industry ordinarily uses the woods native to the immediate area in which the plant is located. Exceptions, particularly in the case of residue, exist when such raw material is shipped over long distances to the manufacturing plants located closer to markets. In North America, there are primarily three areas of intensive forest products operations where raw material availability both in terms of roundwood and manufacturing residue is highly favorable. These areas are the Pacific Coast stretching from British Columbia to central California, the Great Lakes region including the St. Lawrence Seaway area, and the southern and southeastern United States.

Along the Pacific Coast, a number of coniferous woods are used in significant quantities for board manufacture. These include Douglas-fir which is, by far, the most predominant species presently utilized. Other species for particleboard manufacture in this region are Sitka spruce, western hemlock, lodgepole pine, ponderosa pine, western red cedar, white fir, and redwood. A number of marginal hardwoods are being increasingly used in this region for particleboard manufacture. In the Great Lakes region the conifers used in the board industry include red pine, white spruce and black spruce, balsam fir, jack pine, and eastern white pine. The hardwood species in this region currently used in board manufacture include aspen, which is being consumed in substantial quantities. Other hardwoods taken in lesser amounts include basswood, paper birch, and other medium- and low-density species.

In the southern and southeastern United States, the southern pines form the principal coniferous species utilized by the particleboard industry. The industry currently consumes substantial quantities of southern pine residue from stud mills, veneer mills, and other primary manufacturing operations. Although there are some ten species grouped in the southern pine category,[4] the major portion of the raw material used by the particleboard industry consists of loblolly pine, slash pine, shortleaf pine, and longleaf pine. In addition to these pines, there are a large number of southern hardwoods which are increasingly being used both in the form of roundwood and mill residue. Among these, the most important are sweet gum, black tupelo, the various species of oak, sweet and yellow birch, yellow poplar, and others such as hackberry. These are often used in mixtures with the pines or with the less dense hardwoods.

The technological capability of using practically any wood species and many nonwood lignocellulosic materials is available to the particleboard industry. Most often the question of species or raw material mix is settled on economic considerations rather than technological ones, although these two are not always independent. The species question is a minor consideration while the quantity of available raw material at low cost is the overriding factor in establishing new activities. Mixing of species is likely to become more prevalent due to economic necessities particularly in hardwood regions and where the raw material in large part consists of residue from sawmills, plywood plants, and other wood-using industries.

2-3. Types of Raw Material

As mentioned before, there are a number of lignocellulosic materials which could successfully be utilized for particleboard manufacture. These materials, however, must meet the following requirements:

(a) Availability in adequate quantities.
(b) Inexpensive.
(c) Suitable form for board manufacture.
(d) Incurring relatively low cost in handling and storage.

All considered, wood is the most widely used source of raw material.

Wood used in the particleboard industry originates from various sources. In the United States, Stillinger[5] categorizes the wood raw material into undried and dry sources. The undried wood raw material in order of available volume consists of:

(a) Planer shavings generated in surfacing green lumber.
(b) Plywood mill residue, which is normally chipped and results from trimming veneer bolts and veneer clippings.
(c) Sawmill residue, usually consisting of chipped slabs, edgings, and trimmings.
(d) Roundwood which has been reduced into particles.
(e) Sawdust.

The dry raw material consists of:

(a) Planer shavings generated from surfacing kiln dried lumber.
(b) Residue from plywood mills produced after the veneer has been dried.
(c) Residue from dry rough lumber cuttings in furniture and millwork plants.
(d) Dry sawdust.

Currently, planer shavings are the major source of raw material for particleboard manufacture due to their favorable cost and availability even though this source of raw material does not possess the best particle geometry (weak, curled, bulky). Each year, over 350,000 tons of softwood planer shavings are generated in the United States,[6] with a price per ton perhaps one-fourth of that paid for high-quality flake-type particles[7] and as little as 10 percent of that paid for lumber.[8] Research[9] has aimed at finding surfacing techniques which would yield a superior stock surface while at the same time generating higher quality shavings than those commonly produced in industry.

Residue requires less handling and fewer processing steps than solid wood. This implies less labor, capital, and maintenance costs. With sufficient supplies of inexpensive residue available in the United States, there is little economic incentive at this point in time to use forest waste or roundwood. However, from the standpoint of resource conservation, there is a need to find ways of fully utilizing all logging residues (short logs, broken logs, branches, forest thinnings). In addition, the particleboard industry should be made capable of economically using cull timber and noncommercial species normally incapable of bearing the cost of harvesting and transport for primary manufacture. The use of logging residue not only expands the raw material base but it can also be of assistance in proper forest management. Cleaning up such residue frees the land for growing more desirable species. The development and use of mobile chippers makes the use of such residues an economic reality; the development and use of equipment to remove foreign matter will make the raw material technically acceptable. With the anticipated increase in bark inclusion in particleboard products, debarking of such residue may become unnecessary if the bark is sufficiently clean.

In Europe, logging residue does form a source of raw material for the particleboard industry due to the pressures of rising production and limited raw material supply. Extensive examination of branches and early thinnings was taken up by Buschbeck and coworkers[10] and Loycke.[11] The material used by these authors consisted of nontimbered portions of the tree with a diameter of about $1\frac{1}{2}$–3 inches (including bark) at the smaller end. Negative technological attributes of such material include high sapwood (up to 98 percent) and bark content, which is normally expected to result in lower strength boards. However, research by the authors points to no significant deterioration of board properties as particles made from tree branches were added to the core layer of a three-layer, flake-type particleboard (fig. 2–3). This figure shows that even 100 percent branchwood particles in the core produced the same properties as those belonging to a board which possessed none of this type of material.

2–3. *The influence of the proportion of particles from pine branches (added to the core layer of a three-layer particleboard) on indicated strength and dimensional properties.*

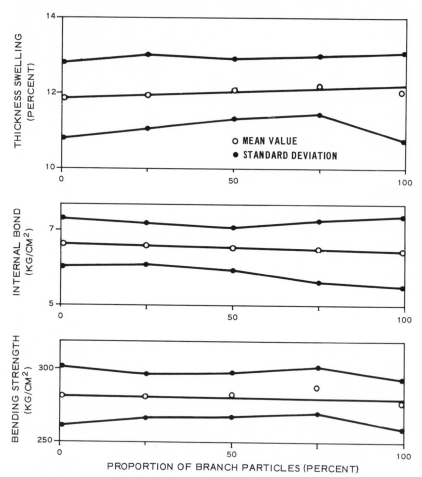

SOURCE: Buschbeck, L., Kehr, E., Scherfke, R., and Jensen, U. 1965. Investigation on the suitability of various wood species and assortments for chipboard manufacture. Part 1: Red beech and pine. Redrawn in modified form by S.I.U. Cartographic Laboratory

Modern resins have also paved the way for the utilization of nonwood lignocellulosic materials—particularly agricultural residues. Bagasse from sugarcane waste is a major fiber source capable of producing acceptable particleboard and other types of fiberbase boards. Bamboos, e.g., *Bambusa arundinacea, B. nutans, B. polymorpha, B. tulda, Dendrocalamus strictus,* and *Oxytenanthera abyssinica,* have shown to be capable of quality board production in the laboratory. Flax shives have been used in Europe for producing particleboard for at least a decade. Other lignocellulosic sources of raw material, e.g., cotton stalks, cereal straw, papyrus, palms, African elephant grass, gum bark, coconut and peanut shells, along with others, hold some possibility for utilization as raw material for board production, particularly in areas where adequate quantities of wood are not available. Also, wastepaper and board and other lignocellulosic plant fiber originating from city refuse are likely to gain significance as board raw material in future decades as the economics and technology of recycling raw material resources are further advanced. None of the nonwood materials, however, is currently free from problems, mostly of an economic nature.

Bagasse, for instance, is often by no means a cheap raw material source. Some of the problems associated with bagasse are:

(a) It has value as fuel to the sugar mills and therefore the price paid for bagasse must cover the fuel costs.
(b) It has to be harvested over a relatively short harvest time (as little as $2\frac{1}{2}$ months in most areas)[12] necessitating large storage facilities so that the board plant may be able to use the material year-round. As a result, costs are generated for collecting, protecting, storing, and handling the material.
(c) There are many varieties of sugarcane each with somewhat different characteristics, presenting problems similar to the processing of mixed wood species. The normal practice for storing and handling bagasse consists of bailing the material and storing it in piles either in the open or under simple roof covers. Special care such as pile separation is exercised to reduce the fire risk. Once the pile is stored, fermentation begins to take place, resulting in a desirable reduction in moisture and sugar contents. Special additives, i.e., boric acid, can be utilized to control mold and fungus growth.

Flax is a cultivated plant with a woody stem approximately three feet high and from $\frac{1}{8}$ to $\frac{1}{4}$ inch in diameter. It may be grown either for its fiber or seed. The fibres, present in the woody stem, are separated from the stem by soaking the stalks in water for a period of time followed by passing them through rollers and beaters. This facilitates the removal of the fibres leaving a lignocellulosic residue called flax shives. Research shortly after World War II in Europe found the shives suitable for particleboard production, leading to the eventual establishment of a number of plants in that continent. Since flax is an agricultural crop, storage problems are similar to those of bagasse. The shives used for board manufacture, being the residues of flax fiber production, result in storage problems at the fiber producing plant. The quantity of flax shives is often limited and is thus only capable of supporting a small particleboard plant —unless the shives are used in mixture with other supplementary materials. In the United States, flax shives have been used for about two decades to produce low-density fiberboard. However, there has been no particular rush to utilize this source in medium-density particleboard primarily due to the availability of adequate quantities of wood.

Cotton stalks form another type of raw material technically capable of producing particleboard similar in properties to that made from wood. The stalks can be chipped, screened, and processed; the particles adhere well with the usual resins. However, the problems associated with harvesting, baling, and transportation make this source of raw material unattractive. Corn stalks and cereal straw also present the same economic difficulties in addition to the fact that these plant sources are not technically easy to use in particleboard manufacture. Corn stalks have to be cleaned of leaves, husks, and pith, and cereal straws deteriorate rapidly in storage (particularly when modern harvesting techniques are used) and present the problems of nodes, entrained dirt, weeds, and the silica content of the straw itself. The use of coconut fiber residue and peanut shells in particleboard particularly as core material is proving practical in certain parts of the world where such raw materials are in plentiful supply and wood is scarce.

2–4. Quality of Raw Material

The quality of raw material affects board properties and is influenced by such factors as species mix, extent of enmeshed impurities, seasonal changes, residue supplier practices, and the particle form of such residue. Apart from availability, the quality of the raw material is a prime consideration in deciding whether or not an economical plant operation is feasible. In localities where the variations in wood species mix and moisture content are large, many different manufacturing problems arise producing difficult operating conditions. When the delivered residue contains significant levels of foreign materials such as tramp metal, cadmium particles, bark, mineral matter, dirt, glass, and stones, not only will the board quality suffer but the impurities also damage the equipment. The installation of segregating facilities take care of such impurities, but the added expense for equipment and operational costs could make the board manufacture uneconomical. The wood-reduction process or the residue generating machinery can introduce contaminants. Oil and oil vapors from machinery or dirty tools will degrade the raw material by interfering with the wetting of the adhesives.

Heartwood and sapwood are not believed to bond similarly; the latter tends to be more absorptive retaining inadequate amounts of adhesive on the particle surface. When the raw material consists of smaller trees and branches, the proportion of sapwood is large. Springwood and summerwood in varying proportions in the raw material could present minor problems for trouble-free operation. Compression wood, incipient rot, and other wood defects develop a bond of low strength. Storing particles in outdoor conditions, although convenient for many mills, encourages industrial and other impurities to mix with the raw material and cause outright losses. A study[13] undertaken in interior British Columbia has indicated that with pulp chips, an average monthly loss of 0.75 percent can be expected with lodgepole pine stored outdoors. The loss was 0.65 percent for white spruce. These losses were observed in an experimental pile in which the maximum temperature was 146°F. over a six-month period. For larger piles, it is not uncommon to attain temperatures of 180°F. or higher; therefore it is probable that economic losses resulting from storage of wood residue outside would be even larger than those discovered. The study just referred to indicates that pine and spruce (and perhaps other species) particles can be stored up to three months with limited losses. Longer storage periods should be avoided, if possible, for the purpose of:

(a) Keeping contamination at the plant site to a minimum.
(b) Limiting wood losses.
(c) Avoiding the possibility of severe chip degradation.
(d) Reducing the risk of particle-pile fires.

It is advisable that particles stored outside should be in long, narrow, low piles to reduce losses.

2–5. Acidity

Wood acidity has little effect on the process of manufacture. It however affects the rate of binder cure when acid-sensitive adhesives (i.e., urea-formaldehyde) are utilized. Most wood species used in the particleboard industry are acidic with some of them, such as oak, having a very low pH. The hardening of urea-formaldehyde adhesives depends on establishing a chemical field which is dependent in part on a certain range of acidity generated by both the wood species and catalysts. In modern operations catalysts are becoming indispensable for practically all plants regardless of the pH of the wood species being utilized since fast curing times are required for modern production speeds.

2–6. Extractives

The presence of extractives in the furnish influences the integrity of particleboard in both negative and positive ways. For instance, extractives can have adverse effects on the setting of adhesives, thereby lowering the strength of the particle-particle bond. Extractives can also be volatilized in the hot press under the influence of high pressure, heat, and moisture and may cause blows severely reducing the internal bond strength. On the positive side, extractives in some cases can impart water resistance to the board.[14] Wax and oil content are problematic in forming sound adhesive bonds, particularly in such species as gum, teak, and Brazilian rosewood. Softwoods and hardwoods may cause blows when their extractive-laden heartwood is present in high proportions in the furnish. In such cases, it is advisable that they be somewhat drier during pressing to reduce the chances for blows. The oaks, with their high tannic acid content, interfere with the durability of the bond, particularly under severe exposure. Larch contains galactans which consist of gummy substances interfering with the proper wetting of the wood by the adhesive. In some cases, such extractives contaminate the adhesive by impairing its cohesive strength. A high content of minerals such as silica cause an unusually excessive rate of wear in cutting knives, saws, and the like, thus constituting a drawback for the particular raw material being used by the mill.

Notes

1. Mean compression ratio may be defined as $c \approx g_w/g_b$ in which g_b signifies the compressed wood density and g_w is the natural, noncompressed wood density.

2. Stegmann, G., and Durst, J. 1964. Beech pressboard.

3. Grigor, A., Fleischer, H., and Mitisor, A. 1968. The static bending strength of particleboard.

4. Other southern pines are: Virginia pine, spruce pine, sand pine, pond pine, and two other minor species.

5. Stillinger, J. R. 1967. Drying principles and problems.

6. Lutz, J. F., Heebink, B. G., Panzer, H. R., Hefty, F. V., and Mergen, A. F. 1969. Surfacing softwood dimension lumber to produce good surfaces and high value flakes.

7. Heebink, B. G. 1964. Improving planer residue for better utilization.

8. Mottet, A. L. 1967. The particle geometry factor in particleboard manufacturing.

9. Lutz, J. F., Heebink, B. G., Panzer, H. R., Hefty, F. V., and Mergen, A. F. 1969. Surfacing softwood dimension lumber to produce good surfaces and high value flakes.

10. Buschbeck, L., Kehr, E., Scherfke, R., and Jensen, U. 1965. Investigation on the suitability of various wood species and assortments for chipboard manufacture. Part 3: Pine brushwood.

11. Loycke, H. J. 1961. The utilization of low-grade wood assortments.

12. In some areas such as Hawaii, fresh bagasse is available year-round making the storage of large quantities of bagasse unnecessary.

13. Anon. 1969. Outside chip storage in B. C. interior.

14. Foster, W. G. 1967. Species variation.

Heebink, B. G. 1967. Wax in particleboards.

Bark in Particleboard

3–1. General Considerations

Utilization of bark constitutes one of the most pressing problems facing the wood particleboard industry. The quantities of bark available in the United States are substantial since approximately 10–15 percent of the volume of every log is bark (table 3–1). In the state of Oregon alone, over 2.5 million tons of bark are produced annually. Dealing with such enormous quantities of bark has created thorny disposal problems, particularly with the more stringent laws regarding air and water pollution now in force. It has become imperative that the industry no longer consider bark a waste. A large portion of the generated bark is now being marketed for a variety of uses ranging from high-quality mulch to poultry bedding. Some of the current markets, however, are seasonal and do not require bark in sufficient quantities to consume the supply available. Thus, the use of bark in board manufacture either in mixture with the wood furnish or alone has constituted a challenging possibility for researchers and industrialists alike. Bark is a low-cost material with relatively insignificant competitive uses by other major industries. This situation does not apply to wood residue: the paper industry is likely to continue competing with the wood particleboard industry for the available supply. There is a strong possibility that this competition will intensify in the future as the demand for raw material is increased.

3–1. *The amount of bark volume (based on wet log volume) for a selected number of species and tree diameters*

Wood Species	Diameter of Tree (in.)	Bark Volume (percent)
spruce	4.2	12.1
	8.7	9.3
true fir	4.5	10.0
	8.4	9.4
white birch	3.9	14.5
	8.3	9.5
yellow birch	4.2	10.1
	9.3	9.3
beech	4.2	6.9
	8.9	6.1
sugar maple	4.1	13.5
	7.8	18.0

SOURCE: Harkin, J. M., and Rowe, J. W. 1969. Bark and its possible uses.

3–2. Bark Structure and Properties

Bark makes up the outer part of the tree stem and branches and consists of a highly complex, heterogeneous material. It contains two layers: a physiologically alive inner layer and a protective outer layer. Most of the bark is made up of the latter, which is relatively inert and is usually dark in color. Outer bark consists of dead phloem cells with the lighter color inner bark composed of accumulated annual layers of phloem. Food material is directed downward by sieve tubes, most of which are inactive except for the portion immediately adjacent to the area of the vascular cambium. The thickness of inner and outer bark varies appreciably among various species. For instance, spruce is characterized by a relatively thin bark, a high proportion of which consists of inner bark, while in redwood and Douglas-fir the bark is generally thick and contains a high percentage of outer bark. Not only do large variations occur among various species, but such variations occur in the same species grown under different conditions.

Bark formation is generated by the vascular cambium whose cells divide, producing xylem on the inner section of the trunk and phloem on the outer. In addition to the sieve tubes referred to above, bark

3–1. *The tons of bark generated per one thousand board feet of gross log scale as a function of diameter classes for three important American softwoods.*

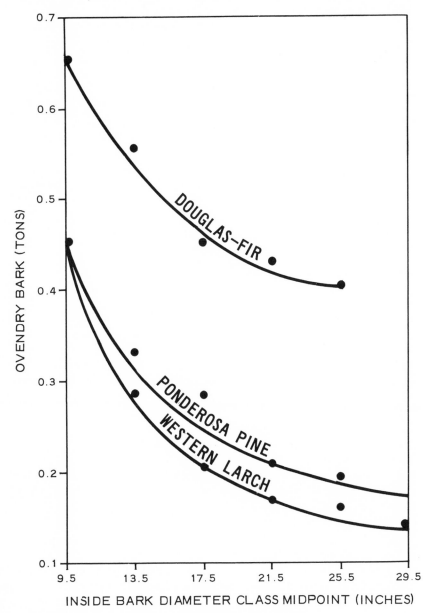

INSIDE BARK DIAMETER CLASS MIDPOINT (INCHES)

SOURCE: Harkin, J. M., and Rowe, J. W. 1969. Bark and its possible uses.
Redrawn in modified form by S.I.U. Cartographic Laboratory

contains phloem parenchyma, bast fibers, and companion cells. Some barks slough off relatively rapidly over time with the result that the bark thickness remains small. In species characterized by this type of bark, the outer appearance remains scaly but smooth. On the other hand, in some species the bark is not cast off, thus accumulating over the years. In such species, the bark may be up to two feet in thickness and usually will have a deeply fissured appearance.

The amount of bark per log also varies appreciably with size. As the log size decreases, the proportion of bark increases. Figure 3–1 shows the relationship between the quantity of bark in tons generated per thousand board feet of gross log scale for various diameter classes of three softwood species. It is noted that a substantially greater quantity of bark is generated when Douglas-fir logs are being processed. The properties of bark, both from chemical and physico-mechanical standpoints, vary widely among various species. Table 3–2, for instance, points to the wide variation in bark specific gravity for a selected list

3–2. *A selected list of the specific gravity of bark of a number of species grown in the United States*

Common Name	Scientific Name	Specific Gravity (oven-dry weight and volume)
California incense-cedar	*Libocedrus decurrens* Torr.	0.269
Englemann spruce	*Picea engelmannii* Parry.	0.797
western white pine	*Pinus monticola* Dougl.	0.625
Douglas-fir	*Pseudotsuga menzisii* (Mirb) Franco.	0.544 0.411
redwood	*Sequoia sempervirens* (D. Don) Endl.	0.459
baldcypress	*Taxodium distichum* (L.) Rich.	0.553
western hemlock	*Tsuga heterophylla* (RAF) SARG.	0.588
mockernut hickory	*Carya tomentosa* Nutt.	0.983
sugar maple	*Acer saccharum* Marsh.	0.686
yellow birch	*Betula alleghaniensis* Britton.	0.741
black walnut	*Juglans nigra* L.	0.378
sweetgum	*Liquidambar styraciflua* L.	0.582
yellow poplar	*Liriodendron tulipifera* L.	0.388
black tupelo	*Nyssa sylvatica* Marsh.	0.546
sycamore	*Platanus occidentalis* L.	0.672
bigtooth aspen	*Populus grandidentata* Michx.	0.664
black cottonwood	*Populus trichocarpa* Torr. & Gray etc.	0.604

SOURCE: Harkin, J. M., and Rowe, J. W. 1969. Bark and its possible uses.

3–2. *Relationship between the strength and the percentage of bark in the furnish for redwood particleboard at three different resin levels.*

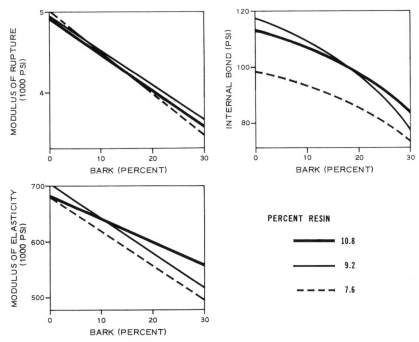

SOURCE: Dost, W. A. 1971. Redwood bark fiber in particleboard.
Redrawn in modified form by S.I.U. Cartographic Laboratory

of species. The literature on the subject is extensive with some excellent bibliographies prepared.[1] The study of bark is an extensive task occupying a significant proportion of research currently being carried out, the elaboration and discussion of which is beyond the scope of this volume. For an introduction to the subject, the reader is referred to the two books[2] cited in the notes at the end of this chapter.

3–3. Wood-Bark Particleboard

With the present level of technological know-how, the use of bark in wood particleboard manufacture is feasible only to a limited extent. This is due to the fact that excessive use of bark (over 10 percent) produces significant adverse effects on strength and dimensional properties of the resulting particleboard. When fibers are used to produce medium density, single-layer particleboard, experiments have shown that bark (particularly hardwood bark) can be tolerated at a maximum of 5 percent without noticeably affecting the board properties.[3] Bark contents of 10 percent or higher in this type of particleboard have been found to reduce seriously strength and dimensional stability. Some barks such as those of the southern pines, cannot be tolerated at all.

Figure 3–2 illustrates how an increasing bark level will bring about reductions in strength properties of particleboard with average density of 0.65 gm/cm[3]. This figure has been derived for three-layer redwood particleboard in which various levels of redwood bark in fibrous form have been added to both face and core layers. Various percentages of average resin content have been used as indicated. It is seen that severe reductions in strength are brought about when bark percentages exceed 10 percent (on weight basis). Equally disappointing is the fact that thickness swell and linear expansion of the board increase when bark in any significant percentage is used (fig. 3–3). Dost[4] reports that other important properties of the board—e.g., surface smoothness—decrease with increasing bark content. In addition to reduction in strength and

3–3. *Deleterious effects of increasing bark percentage in the furnish on water absorption and dimensional stability (water soak) of redwood particleboard.*

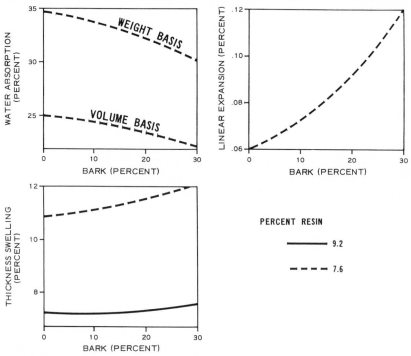

PERCENT RESIN

———— 9.2

– – – – 7.6

SOURCE: Dost, W. A. 1971. Redwood bark fiber in particleboard.
Redrawn in modified form by S.I.U. Cartographic Laboratory

3–4. *Internal bond strength of a three-layer particleboard as an increasing percentage of branch particles (with and without bark) is added to the core furnish.*

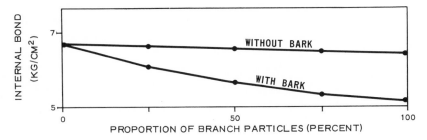

SOURCE: Buschbeck, L., Kehr, E., Scherfke, R., and Jensen, U. 1965. Investigation on the suitability of various wood species and assortments for chipboard manufacture. Part 3: Pine brushwood. Redrawn in modified form by S.I.U. Cartographic Laboratory

dimensional stability, the handling of bark during processing can be problematic. Fibrous bark can ball up during blending and forming. Additionally, the presence of bark creates maintenance problems for wood-reduction equipment and other cutting facilities due to the existence of dirt and other similar foreign materials which become embedded in the bark. A further difficulty arises when variations occur in the proportion of bark in the furnish: The maintenance of a relatively constant furnish pH becomes difficult. The problem of bark usage depends on a number of factors such as species, form of raw material used, type of process, the kind of board made, and the use intended for the produced board. At the present time, the particleboard industry occasionally uses bark primarily in the core layer and even then only in quantities normally occurring in roundwood.

The use of branches without debarking ($1\frac{1}{2}$–3 inches in diameter) has been examined by Buschbeck and coworkers.[5] These authors utilized pine branches which were subsequently reduced into sliverlike particles (t = .014 in., l = .45 in., and w = .07 in.) and used for the core layer of a three-layer particleboard. It was noted that as the percentage of branchwood with bark was increased in the core layer, the strength of the board was somewhat lowered and the thickness swelling increased. The reduction in internal bond strength, however, was significant (fig. 3–4) with the authors attributing this mainly to the presence of increasing quantities of bark in the mix. The appreciable drop in the internal bond strength constitutes a serious setback when unbarked wood is utilized. In such cases, increasing board density and/or core resin content can, in part, compensate for the drop in the internal bond strength as noted in figure 3–2; but such measures may be uneconomical in many cases. In the extrusion process, the quantity of bark that can be accommodated without objectionably affecting the board properties is also limited (to perhaps up to about 10 percent). The demixing problem can likewise cause production difficulties in this process. Limited quantities of bark in the extrusion board is believed by some[6] to act as filler occupying only the cavities in the board.

3–4. All-Bark Particleboard

Burrows[7] made an extensive laboratory study on the technical possibility of using Douglas-fir bark, without the addition of any external binder, for an all-bark particleboard. In this study, bark was reduced to particles by hammermilling with mats being formed as usual (without any prior adhesive blending) utilizing two caul plates. During hot pressing, using a press cycle similar to that employed in the manufacture of wood particleboard, serious panel blowing was encountered due to excessive mat moisture content, overabundance of fine particles in the furnish, and perhaps the existence of extractives in the bark. Controlling the amount of moisture and fines was found to nearly eliminate blowing, at the cost of reduction in board strength and stability. To eradicate panel blowing and yet retain the desirable levels of physical properties of the all-Douglas-fir bark particleboard, experiments by Burrows indicated that a press cycle which includes cooling at the conclusion of the hot pressing period can prove successful. Bark with high moisture and fines content may be utilized, producing boards with acceptably high physical and mechanical properties. However, cooling the hot press for every board load is a prohibitive factor at this time; it is expensive to heat and cool the press frequently during the course of manufacture. In this type of operation, the press will constitute a severe bottleneck, making press time unacceptably long. It is possible to remove various extractives from the bark before use in board manufacture. This extraction generates a number of useful by-products which could make bark utilization less economically forbidding for a board plant.

Using a constant press cycle (heating for ten minutes at 280°F. and cooling with running water for an additional five minutes), Burrows employed three pressure levels (150, 300, and 450 psi), two levels of mat moisture content (12 and 20 percent),[8] two types of overlay along with control (none, veneer, and kraft fiber), and three particle sizes (mesh sizes $-4+10$, $-2+10$, and $-2+20$), to investigate the possibility of producing a commercially acceptable all-bark particleboard. He found that particle sizes chosen did not significantly affect the more important properties of the bark board. On the other hand, such properties in most cases were influenced by press pressure, the type of overlay, and the mat moisture content.

Figure 3–5A shows that modulus of rupture was affected by the type of overlay and the press pressure in a rather predictable manner while mat moisture content, within the range chosen, did not appear to influence this particular property. The boards with veneer overlay provided superior strength due to the veneer surfaces taking up most of the bending load. The modulus of elasticity was also affected in a similar fashion (fig. 3–5B). The screw withdrawal strength of all-Douglas-fir bark boards appeared to compare favorably with commercial wood particleboard products. Figure 3–5C depicts the influence of overlay, mat moisture content and press pressure on the screw withdrawal strength of laboratory-made boards. It is noted that reduction in press pressure generally leads to lower strengths due primarily to a drop in board density. The internal bond strength of the all-bark particleboard, yields satisfactory results when compared to commercial wood particleboard. In laboratory experimentations, the Douglas-fir bark board overlaid with veneer produced lower values of internal bond strength (fig. 3–5D). The lower internal bond values are believed to be, in part, due to the slower heating rate in the center of the veneered board compared to the controls.

The twenty-four-hour immersion test revealed that veneer-overlaid all-bark particleboard generally exhibits a greater degree of thickness instability, although the magnitude is within practically acceptable bounds (fig. 3–6A). The linear stability of Douglas-fir bark particleboard also compares well with commercial board products, particularly when overlaid with veneer or kraft fiber (fig. 3–6B). The extent of linear expansion with kraft overlay was about 0.4–0.5 percent (based on dimension before soaking). With veneer overlay, this was reduced to approximately 0.1 percent. The mat moisture content, before pressing, generally had only minor influence on the linear dimensional stability of the Douglas-fir bark particleboard. The result of water absorption determinations for all-bark boards indicated that veneer-overlaid boards had the poorest performance, absorbing some 9–10 percent (based on weight before soak) after twenty-four-hour immersion in water. It was also noted that when mat moisture content was high (20 percent), a lower water absorption rate was exhibited. Higher press pressures lead to denser boards, resulting in a lower rate of water penetration.

The rate of heat penetration into the interior of bark particleboard is influenced by press temperature, press pressure, and the type of overlay employed. As noted earlier, veneer overlay appears to retard heat penetration compared to non-overlaid controls. Press pressure and mat moisture content have a predictable effect on the rate of heat penetration

3–5. *The influence of the type of overlay on four strength parameters of Douglas-fir bark particleboard as determined over a range of hot press pressures: (A) modulus of rupture, (B) modulus of elasticity, (C) screw withdrawal, and (D) internal bond.*

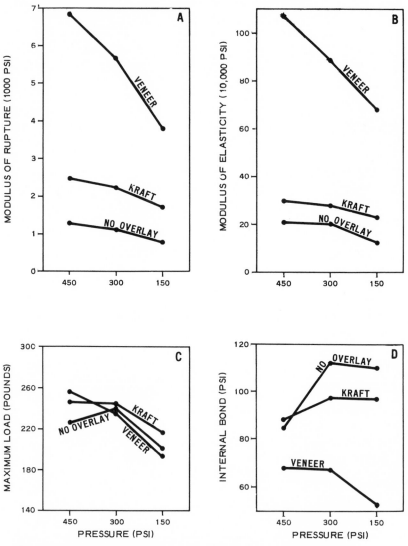

SOURCE: Burrows, C. H. 1960. Particleboard from Douglas-fir bark—without additives. Redrawn in modified form by S.I.U. Cartographic Laboratory

30

3–6. *The influence of the type of overlay on the dimensional changes in Douglas-fir bark particleboard determined over a range of press pressures: (A) thickness swelling and (B) linear expansion. The effect of mat moisture content on the results has also been shown in the latter.*

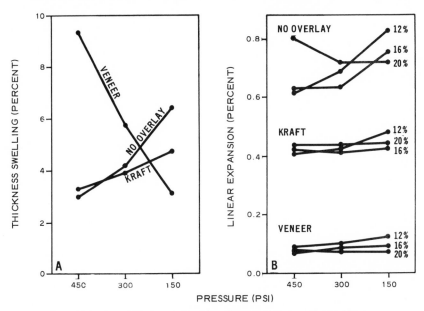

SOURCE: Burrows, C. H. 1960. Particleboard from Douglas-fir bark—without additives.
Redrawn in modified form by S.I.U. Cartographic Laboratory

into the interior of all-Douglas-fir bark board: higher moisture contents or press pressures accelerate heat travel into the board interior after the first two or three minutes have elapsed (fig. 3–7). It is emphasized that the results described above involved Douglas-fir bark under certain circumstances. With commercial softwood and hardwood barks having a wide range of properties, different results (even if the same experimental settings are used) can be expected.

Based on the knowledge available on all-bark particleboard, the following advantages are discernible:

(a) No external adhesives would probably be needed; an obviously significant advantage.
(b) If no adhesives are required by the process, a plant producing bark particleboard will not need adhesive blending or storage facilities.
(c) No after-press cooling station is required since the boards are already cooled when emerging from the press.

3–7. *Interrelationships between press pressure, time, and mat moisture content on heat transfer of Douglas-fir bark board.*

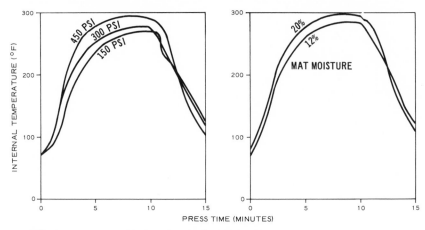

SOURCE: Burrows, C.H. 1960. Particleboard from Douglas-fir bark—without additives.
Redrawn in modified form by S.I.U. Cartographic Laboratory

(d) No sanding equipment is probably called for in the bark particle-board plant since the overlays are often smooth enough not to require in-plant sanding.

The last advantage may not hold if very close thickness tolerances are specified.

The disadvantages include:

(a) A specially-designed press capable of fast heating and cooling is required, indeed a difficult technological task to accomplish with current short press cycles.
(b) Equipment for adhesive application to overlays are needed.
(c) Suitable forming equipment capable of handling bark is probably necessary, particularly if three-layer bark board with fine particles on surfaces is to be produced.
(d) The cost of overlays may make the products economically uncompetitive.

Commercial production of bark particleboard requires a significant degree of economic and technological refinement not presently available. Serious problems exist in raw material collection and processing (adequate quantity, cleanliness, etc.).

Notes

1. Marian, J. E., and Wissing, A. 1956–1957. The utilization of bark: Index to bark literature.

Roth, L. 1960. Structure, extractives and utilization of bark.

Roth, L. 1968. Structure, extractives and utilization of bark. Suppl. 1.

Roth, L., and Weiner, J. 1967. Barkers and barking of pulpwood. Suppl. 1.

Roth, L., Saeger, G., Lynch, F. J., and Weiner, J. 1960. Barkers and barking of pulpwood.

2. Jensen, W., Tremer, K. E., Sierila, R., and Wartiovaara, V. 1963. The chemistry of bark.

Srivastava, L. M. 1964. Anatomy, chemistry and physiology of bark.

3. Raddin, H. A. 1967. High frequency pressing and medium density board.

4. Dost, W. A. 1971. Redwood bark fiber in particleboard.

5. Buschbeck, L., Kehr, E., Scherfke, R., and Jensen, U. 1965. Investigation on the suitability of various wood species and assortments for chipboard manufacture. Part 3: Pine brushwood.

6. Connelly, T. J. 1955. The Kreibaum process of extruded core board.

7. Burrows, C. H. 1960. Particleboard from Douglas-fir bark.

8. The author actually used an additional level of moisture content—16 percent—with the results generally falling between the 12 and 20 percent levels.

Furnish Moisture

The effect of furnish moisture content on the particleboard process and properties is highly significant. The moisture content of the furnish originates from four sources as follows:

(a) The original or retained water remaining in the wood particles after they have undergone drying. The amount of moisture attributed to this source varies from plant to plant but it generally ranges between 3 to 6 percent based on the oven-dry weight of the wood particles.

(b) Water introduced by adhesives. Most adhesives used in the particleboard process are aqueous dispersions of high molecular weight polymeric materials. With the application of the adhesive (which may have as much as 35 to 60 percent water content), an additional amount of water is introduced into the wood particles. Part of this water, however, is evaporated during the flight of the adhesive droplets in the spraying process and after it has struck the particle. The remaining portion of the water eventually diffuses into the particles during the blending, pressing, and curing period. The water from this source is capable of elevating the furnish moisture content from 3 to 5 percent.

(c) The moisture added to the mat by the urea-formaldehyde condensation process (which generates water as its by-product). The amount of water added in this manner is not significant. When using 6 percent

35

resin solids, for instance, this can add a maximum of 0.9 percent to the overall moisture content of the furnish.[1]

(d) The water added by spraying the mat surface (a frequent European practice) to obtain certain desirable end results such as smoother board surfaces and a more rapid heat transfer to the board core (thereby allowing shorter press times). This source, too, does not contribute a great deal to the overall moisture content of the furnish (approximately 1 percent).

Resultant Mat Moisture Content: The resultant moisture content of the furnish from the various sources mentioned above can be calculated. For the purpose of simplicity and due to lack of significance, the moisture added to the mat by the urea condensation process is ignored. In fact, we shall first take into consideration the two most important sources of moisture, namely, the wood particle moisture and the water added during resin and additive blending. In such a case, therefore, we can write:[2]

$$M_R = \frac{W_w + W_b}{W_o + W_r} \qquad (4-1)$$

in which W_w is the weight of water contained in wood particles, W_b represents the weight of water added during blending, W_o signifies the oven-dry weight of the wood particles and W_r is the weight of resin solids. M_R in equation 4-1 thus is the fractional moisture content of the furnish when its two most significant moisture components are taken into account. If we let $M_w = \dfrac{W_w}{W_o}$, $k = \dfrac{W_r}{W_o}$, and $M_r = \dfrac{W_b}{W_r}$, then equation 4-1 can be rewritten as:

$$M_R = \frac{M_w}{(1+k)} + \frac{kM_r}{(1+k)} = \frac{M_w + kM_r}{(1+k)}. \qquad (4-2)$$

Equation 4-2 indicates that by knowing three parameters M_w, M_r, and k, we can arrive at the resultant moisture content (expressed as a fraction of unity) of the furnish if no spraying of water on mat surfaces takes place. However, if water is being sprayed on the mat surfaces in a given plant, equation 4-2 will then contain an additional term taking into account the sprayed water:

$$M_R = \frac{M_w}{(1+k)} + \frac{kM_r}{(1+k)} + \frac{W_s}{W_R}. \qquad (4-3)$$

In this equation, W_s is the weight of the sprayed water and W_R is a parameter expressed as follows:

$$W_R = W_o\,(1+k) = W_o + W_r.$$

Equation 4–3 can now be written as:

$$M_R = \frac{M_w}{(1+k)} + \frac{kM_r}{(1+k)} + \frac{W_s}{(1+k)W_o}. \qquad (4\text{–}4)$$

Equation 4–4, however, applies to a single-layer mat intended for the manufacture of single-layer particleboard. In the case of three-layer boards with faces and core having differing levels of moisture content, the resultant moisture, M'_R, can be calculated as follows:

$$M'_R = \frac{\Sigma W_m}{\Sigma W_o} \qquad (4\text{–}5)$$

in which W_m represents the weight of water present (from various sources) in face and core layers and W_o is the weight of dry wood in face and core layers. Equation 4–5 may be expanded to read:

$$M'_R = \frac{M_c g_c(t-2s) + M_f g_f \cdot 2s}{g_c(t-2s) + g_f \cdot 2s}. \qquad (4\text{–}6)$$

In this equation, M_c is the core moisture, M_f the face moisture, g_c and g_f the core and face densities respectively (oven-dry basis), t the overall board thickness, and s the thickness of a single face layer (fig. 4–1). Dividing the numerator and the denominator of equation 4–6 by t and letting $\lambda = \dfrac{2s}{t}$ (shelling ratio), we obtain:

$$M'_R = \frac{M_c g_c(1-\lambda) + M_f g_f \lambda}{g_c(1-\lambda) + g_f \lambda}. \qquad (4\text{–}7)$$

4–1. Schematic make-up of a three-layer board with face thickness s and core thickness t–2s.

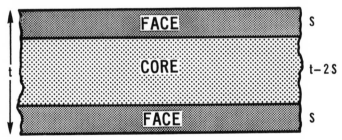

in which $g_R = g_c(1-\lambda) + g_f\lambda$ represents the resultant density of the three-layer board. Now, if we let

$$u = \frac{\lambda g_f}{(1-\lambda)g_c} \tag{4-8}$$

then equation 4–7 can be written as:

$$M'_R = \frac{M_c}{(1+u)} + \frac{uM_f}{(1+u)}. \tag{4-9}$$

Taking into account the water sprayed on surfaces (if this is being practiced), then an additional term must be added to equation 4–9 as follows:

$$M'_R = \frac{M_c}{(1+u)} + \frac{uM_f}{(1+u)} + \frac{W_s}{W_R}. \tag{4-10}$$

W_s and W_R represent the same parameters as those described for equation 4–3. It is noted that a basic similarity between equations 4–3 and 4–10 for single- and three-layer boards exists. In equation 4–10, M_c and M_f represent the moisture contents which include:

(a) The wood particle moisture.
(b) Water added during blending.
(c) Water sprayed on surfaces (if practiced).

Thus, if M_c and M_f are expanded into these moisture components utilizing equation 4–3, we obtain:

$$M'_R = \frac{u}{(1+u)} \cdot \frac{M_{wf}}{(1+k_f)} + \frac{M_{wc}}{(1+u)(1+k_c)}$$
$$+ \frac{u}{(1+u)} \cdot \frac{k_f M_r}{(1+k_f)} + \frac{k_c M_r}{(1+k_c)(1+u)} + \frac{W_s}{W_R}. \tag{4-11}$$

In this equation, the first two terms account for the moisture contributed by the wood particles themselves, the third and fourth terms take into consideration the moisture added during blending, and the last term accounts for the sprayed-on water. The indices f and c refer to parameters belonging to face and core layers respectively. Equation 4–11, therefore, indicates that by knowing the values of wood moisture M_w, solid resin content k, the weight parameter u (see equation 4–8), and the weight of the sprayed-on water, the resultant mat moisture M'_R can be calculated. It is interesting to note that equations 4–3 and 4–11 show that spraying the board surface with water raises the overall mat moisture content to the extent noted earlier.

4–2. Shear strength-moisture content relationships for veneers before they are made into plywood.

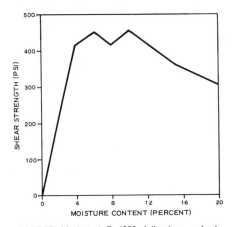

SOURCE: Marian, J. E. 1958. Adhesives and adhesion problems in particleboard production.
Redrawn in modified form by S.I.U. Cartographic Laboratory

4–2. Furnish Moisture Versus Technological Properties

Furnish moisture exerts much influence both on the manufacture and the properties of particleboard. Too high or too low levels of furnish moisture result in troublesome operation and produce a poor quality board. The optimum moisture content depends on many factors such as the nature of the process, particle geometry, and wood density, among others, making generalizations difficult. Marian,[3] based on his studies with veneer, contends that maximum board strength occurs when the moisture content ranges between 8 to 12 percent at the particle-particle interface (fig. 4–2). Kehr and coworker[4] experimented with a three-layer spruce particleboard having uniform moisture content throughout the various mat layers (with the exception of spraying 100 gm/m^2 of water on mat surfaces).[5] Maximum bending and internal bond strengths were obtained when the furnish moisture ranged from 11 to 13.5 percent (table 4–1). Maku and coworkers,[6] using press times of less than five minutes, showed that shear strength was maximized when the furnish moisture ranged between 14 to 18.5 percent (fig. 4–3). Generally, it is advisable that boards made from low-density woods possess a relatively lower moisture content compared to boards made from denser woods. The generated steam in the mat, during hot pressing, travels away from

4–1. *Properties of three-layer particleboards made from spruce furnish with a range of moisture contents (after resin blending) using a press pressure of 13.5 kg/cm² and a press time of seven minutes*

Property	Number of Specimens	Furnish Moisture Content—percent				
		8.5	*11.0*	*13.5*	*16.0*	*18.5*
Board thickness (mm)	12	20.3	20.3	20.6	21.0	21.2
Density (gm/cm³)	12	0.61	0.60	0.61	0.50	0.59
Bending strength (kg/cm²)	15	236	264	248	244	190
Internal bond (kg/cm²)	18	3.8	5.5	6.5	3.4	1.7
Press closing time (min.)	3	1.7	1.7	1.4	1.4	1.3

SOURCE: Kehr, E., and Schoelzel, S. 1968. Studies on the pressure diagram for the manufacture of particleboard. Part 2: Effect of chip moisture, mold closing time and molding pressure on the sealing characteristics in the hot pressing of particleboard.

the hot platens toward the center of the board. The steam also diffuses from the center of the board toward the edges. To make a board of given density and volume using a low-density wood species will require a large number of particles (which are inherently pliable). This is conducive to compaction, which can, to a significant extent, prevent the

4–3. *The effects of furnish moisture content and press time on the shear strength of a 19-mm-thick particleboard.*

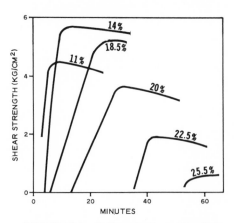

SOURCE: Maku, T., Hamada, R., and Sasaki, H. 1959. Studies of particleboard. Part 4: Temperature and moisture distribution in particleboard during hot pressing.

Redrawn in modified form by S.I.U. Cartographic Laboratory

4–4. *Press closing for mats of different moisture contents over the last 10 mm just before board thickness is reached.*

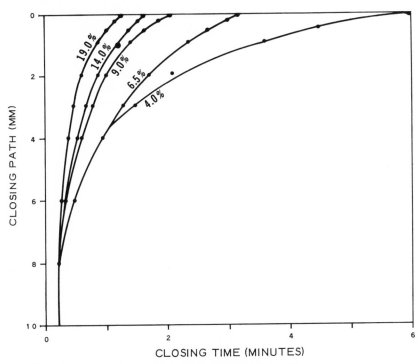

SOURCE: Kehr, E., and Schoelzel, S. 1968. Studies on the pressure diagram for the manufacture of particleboard. Part 2: Effect of chip moisture, mold closing time and molding pressure on the sealing characteristics in the hot pressing of particleboard.
Redrawn in modified form by S.I.U. Cartographic Laboratory

escape of a large volume of water within a short time, a requisite in short press cycles. Thus, excessive moisture in mats made up of low-density particles is likely to lead to low board strength or, in extreme cases, to result in blows and blisters.

Higher levels of mat moisture make wood particles more pliable regardless of their original density. The added moisture content helps produce a more compressible mat when it reaches the hot press. Compressing, with constant pressure, the last 10 mm of the mat before reaching the stops takes less time as the mat moisture content is increased within a 4 to 19 percent range (see fig. 4–4).[7] Excessive mat moisture generally leads to longer total press times due to the moisture's retarding action on the curing of the adhesive. It is responsible for core delamination and occasionally delaminations closer to the board surfaces during hot pressing. The latter, according to Maku[8] is due to occurrence of two moisture maxima at about the first and the last quarter of board thickness (fig. 4–5). This phenomenon results in the accumulation of moisture at these locations washing away the adhesive[9] and generally interfering with the bonding action.

Under certain unusual circumstances, increased moisture content of furnish is necessary in order to obtain boards of adequate properties. Soviet research[10] dealing with indigenous larch sawdust using no external adhesives (utilizing the *in situ* gum as the binder) showed that to make boards successfully from a furnish of this type, the moisture content should be well in excess of 20 percent employing long press times.[11] In this particular study, a furnish moisture content of 28 percent was found to be optimal for obtaining desirable board properties when sawdust of mixed particle sizes was utilized (table 4–2). Decreasing or increasing the moisture content from 28 percent, resulted in reduced thickness stability. The very high levels of furnish moisture content in such cases are a limiting factor as to type of equipment used: when radio-frequency heating systems are employed, such moisture levels can oversupply the board with heat energy and result in a prohibitively high consumption of electricity.

Too low a mat moisture content is also a problem in the particleboard process. Low mat moisture slows heat transfer from the surfaces to the core. Moisture converted into superheated steam reaches the board core quickly, traveling around the particles and generally aiding heat transfer to the interior regions of the board. The fast transfer of heat from the mat faces to the core is essential in short press cycles. On the contrary, too low a mat moisture can lead to a nonuniform panel density and an

4–5. *Moisture distribution over consolidated mat thickness as the press time progresses in a three-layer, 19-mm-thick particleboard.*

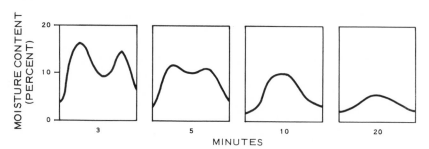

SOURCE: Maku, T., Hamada, R., and Sasaki, H. 1959. Studies of particleboard. Part 4: Temperature and moisture distribution in particleboard during hot pressing.
Redrawn in modified form by S.I.U. Cartographic Laboratory

4–2. *The influence of furnish moisture on product properties when sawdust particles are consolidated into panels without the use of external binders*

Furnish Moisture Content (percent)	Board Moisture Content (percent)	Board Density (gm/cm³)	Bending Strength (kg/cm²)	Water Absorption in 24 hour soak (percent)	Thickness Swelling in 24 hour soak (percent)
24	4.9	1.01	195	disintegrated	disintegrated
27	4.7	1.02	200	disintegrated	disintegrated
28	5.1	1.07	190	18.9	9.6
30	5.0	1.07	190	21.6	15.7
33	5.0	1.06	207	29.5	18.8
35	5.6	1.06	187	30.6	19.2

SOURCE: Vakhrusheva, I. A., and Petri, V. N. 1964. Use of comminuted larch wood for making plastics without binders.

unacceptably rough surface.[12] In addition, low moisture content can lead to poor surface wetting characteristics for the particles, thereby inhibiting resin flow and transfer.[13] Some also believe that overabsorption of resin by excessively dry particles deprive the particle-particle contact area of sufficient binder. Perhaps the most important drawback associated with dry mat is the resultant reduction in the total contact area in the mat structure due to lack of pliability (particularly if high-density wood species are used). A logical outcome for quality board production has been the practice of creating differential moisture contents with the surface layers possessing higher levels of moisture compared to that for the core. Moisture differentiation can result in a number of advantages during the hot pressing operation by:

(a) Rapidly consolidating the mat surfaces into a dense, strong layer as table 4–3 indicates. This phenomenon, however, somewhat lowers the core density by absorbing more material into the board faces. As an increasing amount of water is sprayed on the mat surfaces, a greater surface density, and a wider gap between the face and core layer densities result.
(b) Providing a more rapid heat transfer making a more complete resin cure in the core layer possible with short press cycles.
(c) Preventing precure at board surfaces: higher moisture contents act as cure retardants.[14]
(d) Generating smooth board surfaces due to highly pliable particles being placed in surface layers.

4–6. *The effect of surface water spray (on fibrous mats) upon strength, dimensional and technological properties.*

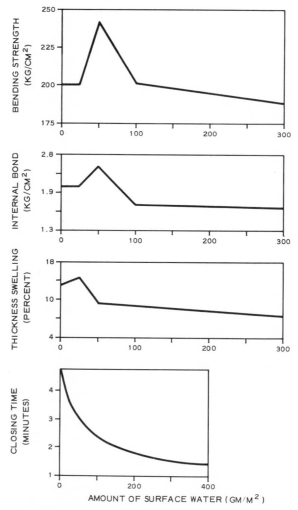

SOURCES: (Upper three) Hinselmann, D., Jentsch, S., and Drechsler, W. 1965. Investigations of manufacturing medium density fiberboard according to the fiber drying process.

(Lower) Kehr, E., and Schoelzel, S. 1966. The investigation of pressing conditions in the manufacture of particleboard.

Redrawn in modified form by S.I.U. Cartographic Laboratory

In order to create a moisture differential, either surface particles with higher moisture contents (compared to the core layer) are used or, just prior to hot pressing, the mat surface is sprayed with a given quantity of water per unit surface area. An optimum exists in the amount of water sprayed, depending on such factors as particle shape and size, and mat moisture content. For instance, figure 4–6 shows the result for a medium-density (690 kg/m^3) particleboard made with fibers.[15] This figure points to an optimum spray of 50 gm/m^2 at which bending and internal bond strengths are maximized while thickness swelling is kept low. With a water spray of 50 gm/m^2 optimum density differentiation occurs between the board's surface layers and the core. As the amount of sprayed water is increased beyond this level, the relative core density becomes too low, causing shear failure in the core layer upon loading. In this study, the internal bond strength was maximized with a water spray of 50 gm/m^2, even though the core density was not at its maximum value (see table 4–3). This is believed to be due to better adhesion opportunities resulting from improved adhesive curing and more pliable particles.[16] Greater amount of water spray results not only in a reduction in core density, but it probably also causes adhesive leaching, thereby leaving the particle-particle interface without adequate binder. One feature of water spraying is its influence on shortening closing time, in spite of the fact that it affects overall mat moisture only slightly (fig. 4–6).

4–3. Monitoring Furnish Moisture

Due to the importance of controlling furnish moisture, constant attention should be focused on this vital parameter if problems associated with moisture are to be corrected promptly. It is advisable to employ means which provide continuous and reliable data at critical stations throughout the entire manufacturing process. Moisture should be checked at the raw material receiving stations, before processing, before and after drying, before and after blending, and just prior to pressing. It is also advisable that monitors keep watch on the moisture content of the consolidated board as it emerges from the hot press, after sanding and before shipping. Wide variations can occur at various stages of manufacture due to the malfunctioning of various processing equipment or variations in moisture content of the raw material received from different sources or locations. Any of a number of reasons; e.g., changes in sapwood/heartwood content, introduction of a new or related species, changes in the ratio of species mixture, and modifications and changes in the practice of manufacture by residue suppliers can influence the moisture content of the raw material.

4–3. *The density of surface and core layers for a particleboard panel with the average density of 690 kg/m³ when different amounts of water are sprayed on the mat surfaces before hot pressing. The press closing rate: 50 mm per second*

Amount of Water Sprayed (gm/m²)	Density (kg/m³)	
	Surface	*Core*
0	700	590
50	900	565
100	980	550
200	1020	545
300	1040	540

SOURCE: Hinselmann, D, Jentsch, S., and Drechsler, W. 1965. Investigations of manufacturing medium density fiber board according to the fiber drying process.

Brumbaugh[17] lists four known basic approaches to measuring moisture continuously in a particleboard plant:

(a) Monitoring moisture by measuring the dielectric constant of wood as it varies with moisture.

(b) Measuring the rate of microwave absorption as it varies with changes in moisture.

(c) Measuring the changes in the relative humidity in the air immediately surrounding the wood furnish.

(d) Making measurements in absorption of a light beam in the infrared range as moisture content changes.

None of the systems listed above, however, have performed flawlessly. They either are disturbed by such variables as temperature, density, species, and bulk density, or the instruments do not remain reliably drift-free without frequent adjustment. Refinement of current instruments designed for continuous monitoring of moisture content is needed. This refinement should lead to improvements in the cumbersome practice of discrete measurements.

Notes

1. Crawford, R. J. 1967. Pressing techniques, problems, and variables.
2. Keylwerth, R. 1958. On the mechanics of multi-layer particleboard.
3. Marian, J. E. 1958. Adhesives and adhesion problems in particleboard production.

4. Kehr, E., and Schoelzel, S. 1968. Studies on the pressure diagram for the manufacture of particleboard. Part 2: Effect of chip moisture, mold closing time and molding pressure on the sealing characteristics in the hot pressing of particleboard.

5. The boards were made with urea-formaldehyde resin using a press pressure of 13.5 kg/cm^2.

6. Maku, T., Hamada, R., and Sasaki, H. 1959. Studies of particleboard. Part 4: Temperature and moisture distribution in particleboard during hot pressing. Part 5: Influence of press time on thickness and cleavage strength of particleboard.

7. Kehr, E., and Schoelzel, S. 1968. Studies on the pressure diagram for the manufacture of particleboard. Part 2: Effect of chip moisture, mold closing time and molding pressure on the sealing characteristics in the hot pressing of particleboard.

8. Maku, T., Hamada, R., and Sasaki, H. 1959. Studies of particleboard. Part 4: Temperature and moisture distribution in particleboard during hot pressing. Part 5: Influence of press time on thickness and cleavage strength of particleboard.

9. Bryant, L. H., and Humphreys, F. R. 1958. Building boards with sawmill waste.

Iwashita, M., Matsuda, T., and Ishihara, S. 1960. Studies on particleboard. Part 1: On the curing condition, specially moisture content of wooden particles. Part 3: Studies on the pressing.

10. Vakhrusheva, I. A., and Petri, V. N. 1964. Use of comminuted larch wood for making plastics without binders.

11. A conventionally heated press was used which applied a pressure of 25 kg/cm^2 at 170°C. with a press time of one minute per each mm. of thickness of the particleboard.

12. Anon. 1969. Particleboard manufacture.

13. Clad, W. 1967. Phenolic formaldehyde condensates as adhesives for particleboard manufacture.

Hemming, C. B. 1962. Wood gluing.

14. Stillinger, J. R. 1967. Drying principles and problems.

15. Fibers were conditioned to 5 percent moisture content, using 9 percent urea-formaldehyde resin (with hardener) and employing a press time of 2–5 minutes. Temperature of 155–200°C. was employed in pressing.

16. Buschbeck, L., and Kehr, E. 1960. Investigations of shortening of compression time in hot compression of particleboards.

Hinselmann, D., Jentsch, S., and Drechsler, W. 1965. Investigations of manufacturing medium density fiberboard according to the fiber drying process.

17. Brumbaugh, J. I. 1969. Plant experiences in controlling moisture.

Particle Geometry

5-1. General Considerations

Particle geometry (shape and size) is a prime consideration affecting both the board's important properties and its manufacturing process. Indeed, the performance of particleboard is, in large part, the reflection of particle characteristics.[1] Mechanical strength, i.e., bending, tension parallel and perpendicular to the board surface, screw and nail holding, is an important property of the board and is greatly affected by particle geometry. Users of particleboard, particularly furniture manufacturers, require smooth surfaces and tight edges when the board is to be overlaid with thin veneers or other types of overlays. Particle geometry affects the face and edge appearance significantly. Thin and small particles with their pliability and gap-filling ability generate gap-free surfaces. In recent years, most commercial operations have utilized fine, dustlike particles and pressure-refined fibers on board surfaces or throughout to achieve exceptionally smooth surfaces. By influencing surface smoothness, particle geometry indirectly influences the finishing, gluing, and overlaying characteristics of particleboard. Particle pliability not only has direct influence on surface and edge appearance, it also to a large extent determines the degree of particle-particle contact and thus affects board strength.

The response of particleboard to water is a vital property affecting such board characteristics as water (liquid or vapor) absorption, thickness

49

and linear stability, strength and surface smoothness. Particle geometry controls this response appreciably. Further, the behavior of particleboard to machining (i.e., sawing, routing, shaping, planing, and sanding) is also affected by the type of particle used in manufacturing the product.

In drying, the shape and size of particles has a direct influence on the amount of heat required for the particles to reach a certain level of moisture content. An increase of particle size of any specific shape, necessitates a greater amount of heat to remove a given weight of water. Smaller particles have a larger exposed surface area resulting in faster water evaporation. Wide variations in particle size, therefore, can result in nonuniform drying with smaller particles tending to overdry, while larger ones will be underdried when compared to the desired level of moisture content.

5–2. Particle Types

The term *particle* is a generic term applied to all lignocellulosic elements from which particleboard is made. Particles are produced either through the action of a hammermill, generating small pieces of indistinct geometry, or by a set of knives as in flakers, which produce engineered particles of predetermined dimension. Confusion has resulted in conjunction with terms used to describe the elements from which particleboard is made. In the literature, the use of *coreboard*, *chipboard*, and *flakeboard*, among others, is frequent when referring to various types of particleboard. In recent years, the general acceptance of the word *particle* when referring to the particleboard's basic lignocellulosic component has been a welcomed improvement over the terminological confusion of the past. Presently, the industry is using the term *particleboard* to describe a family of products made with the various types of particles. Particle types (fig. 5–1) used in the production of particleboard can be as coarse as pulp chips or as fine as sander dust. Some particles are by-products of woodworking operations while others are principally generated for the specific purpose of particleboard manufacture.

Flakes[2] are particles of predetermined dimension and are produced by the action of knives cutting across the grain (either radially, tangentially, or at an angle in between). They are flat, thin particles whose relatively uniform thickness usually ranges from 0.008 to 0.016 inch. Flake length may range from less than an inch to over 4 inches and width may vary from a small fraction of an inch to $\frac{3}{4}$ inch or more. Heebink and coworkers[3] contend that the near-optimum flake for particleboard production is approximately 0.015 inch thick by one inch long and possesses random

5–1. *Various types of particles used in wood particleboard. Top row (left to right): fibers, chips, shavings, and flakes. Bottom row (left to right): excelsior, strands, slivers, and granules.*

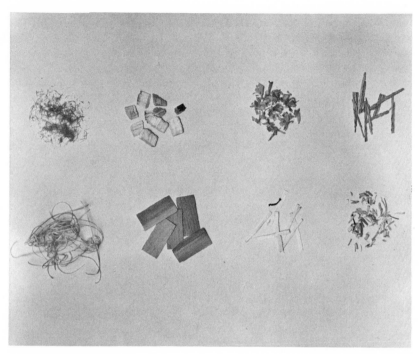

Photograph by Rip Stokes

width. Much research has indicated that this type of slenderness ratio is capable of producing the strongest, stiffest, and most linearly stable boards. The optimum flake also has the grain direction parallel to its long sides and edges; otherwise, deviation will weaken the flake and, subsequently, the board of which it is made. Superlong flakes (two inches or more) are more desirable from the standpoint of producing high-strength boards. However, such flake size is difficult to felt under most industrial conditions. The advantages offered by flakes are partially offset by the fact that they can only be produced from certain forms of raw material and that flakes delivered to a given point in a production system incur higher costs than any other particle type. Flakes, which basically resemble small pieces of thin veneer, can be produced from a limited range of raw material forms; i.e., roundwood, green pulp chips, small size mill waste, and certain types of residue.

Shavings are thin particles of wood produced in a number of wood-working operations utilizing rotary cutter-heads to process the stock. The production of shavings is incidental to such operations. Shavings form the major by-product of such mills as stud mills and furniture plants, where large quantities of wood must be surfaced. Shavings are usually cut along the grain with a thick edge on one side and a feathered edge on the other and are often curled to varying degrees. They produce a particleboard which is inferior in strength, stiffness, and linear dimensional stability compared to boards made from flakes. This is due to the fact that particle configuration in shavings has a number of characteristics which contribute to the manufacture of inferior boards:[4]

(a) They are curled and, therefore, if not broken up, may fold over when consolidated into particleboard.
(b) They are structurally weak.
(c) The thick end often exceeds the optimum thickness required to manufacture high-quality particleboard.

On the positive side, the feathered edges in these particles bring about good felting, good thickness stability, and high internal bond strength. Shavings are generally available in adequate quantities at low cost. The cost factor has made shavings the most important single source of particle raw material, providing the United States with over $\frac{3}{4}$ of its total supply. Shavings can be used as received although much is refined before use. Dust and oversized particles are segregated by screening before entering the process. Oversized particles are usually reduced in size and returned to the plant.

Chips are solid chunks severed from solid blocks of wood by machines that generate axlike cuts. Chips are commonly used in the paper industry. Their size, varying to some extent, measures about 0.5–1 inch in length and width with their height being somewhat less. In the particleboard industry, chips are normally reprocessed into smaller particles before use. Research at the United States Forest Products Laboratory in Madison, Wisconsin, has been directed toward utilizing the large chips (for the core layer), without reduction, in certain experimental particleboards intended for construction use. Chips are generally produced from round-wood although, during the last decade, much of the sawmill and furniture solid wood waste (e.g., trims and slabs) has also been reduced to chips for further utilization.

Fibers used in the fabrication of particleboard consist of a conglomeration of individual and aggregate fibers much similar to those used in making hardboard. It is becoming difficult to distinguish particleboard from hardboard. Pressure-refined fibers are being used either alone or in combination with other particle types to produce a homogeneous or a layered board structure.

Pressure-refined fibers have become increasingly popular in the field of particleboard manufacture. In the United States and in Europe, all-fiber panels in the density range of 36–48 pcf are being commonly produced. Thicknesses in these panels usually range from 15–25 mm. Decisively better surface appearance can be obtained with fibers even when comparison is made with a particleboard which has fines on surfaces. This advantage has brought about a significant demand for fibrous particleboard especially in applications where the board is overlaid with extra-thin veneers. Particleboard made from fibers has uniform properties with supersmooth surfaces, tight edges, an acceptably low water absorption, and is dimensionally stable. Fibers are produced from green or cooked chips utilizing single or double revolving disk mills which produce fibers and fiber bundles. Sometimes, the term *semi-fiber* is used to describe the latter. Fibers are obtained from hardwoods and softwoods; suitable fiber has also been produced from nonwood lignocellulosic sources; i.e., bagasse, esparto, and lemon grass. Fibers have also been obtained from such low-grade residues as sawdust.

Granules, similar to chips, are also chunks of wood severed by axlike cuts. Granules, as defined here, are considerably smaller in size (compared to chips) with their dimensions measuring only a small fraction of an inch. Granules are by-products of certain operations such as sanding or sawing while chips are principally produced and are not incidental to any other operation. Some types of granules, often referred to as "fines," are used in many plants for deposition on the board's surface layers.

Slivers are rectangular or square in cross section with their length measuring at least four times that of their thickness. In slivers (also referred to as splinters) the grain direction runs parallel to their length. Slivers can be pictured as broken match sticks. They vary in dimension to a great extent. Usually they have a length of $\frac{5}{8}$-inch or shorter with their thickness being $\frac{1}{4}$-inch or less. Slivers are produced by hammer-milling mill waste and are considered coarse particles whose configurations are not highly desirable. When used alone, they will not generate a particleboard of high quality. On the other hand, if combined with such particle types as flakes or fibers, a layered board of adequate properties including smooth surfaces can be manufactured.

Strands are defined by some as relatively long shavings consisting of flat, long bundles of fibers with parallel surfaces. Others have referred to long flakes (over $1\frac{1}{4}$ inch in length) as strands. Strands enjoying favor in earlier years of particleboard development, have fallen out of favor due to their difficult-to-felt, extra-long lengths and their high cost of production.

Excelsior, also known as wood wool, is composed of long, curly particles of slender configuration used in some particle panels bonded either with high polymeric adhesives or inorganic binders. One product made with excelsior-type particles and portland cement, is referred to as "excelsior board." Utilization of excelsior has not been confined to board manufacture. It has also been extensively utilized as packing material for fragile goods.

The reduction of wood raw material into a given particle type depends on the form of raw material a plant receives. For instance, when the available raw material is in round form or large green solid waste, it can be reduced to any particle type and size desired. On the other hand, when residue of small particle configuration is being received, there is only a limited variety of particle types which can be generated. In general, the following list indicates the highest (first) to lowest (last) amount of freedom available in selecting the kind of particle type desired:

1. Undried roundwood of various diameter sizes
2. Veneer peeler cores
3. Cull lumber
4. Undried lumber end trims
5. Dry lumber end trims
6. Undried veneer log round-up
7. Undried veneer clippings
8. Undried pulp chips
9. Undried planer shavings
10. Small dry trims from woodworking operations
11. Dry planer shavings
12. Undried sawdust from sawmills
13. Planing mill and similar sawdust
14. Miscellaneous woodworking residues (routing, boring, shaping, etc.)
15. Plywood sanderdust
16. Miscellaneous woodworking sanderdust

5-3. Slenderness and Flatness Ratios

The principal dimensional elements of particles of distinct geometry are length, l, width, w, and thickness, t. Of these, the ratio of length over thickness, called slenderness ratio, s, is a highly important parameter. Thus, slenderness ratio can be described as:

$$s = \frac{l}{t}. \tag{5-1a}$$

In this equation, s is independent of width, as noted. Slenderness ratio is related to an array of vital board characteristics such as contact area in the mat, mechanical properties of the finished board, and the consumption of the binder per a given set of board properties.[5] For lower s values, a greater quantity of resin per unit surface area of particles is consumed as table 5–1 shows. Core particles with normally lower s values require a greater quantity of resin per unit surface area compared to the surface particles. The reason for this is threefold:

(a) The lateral surfaces and ends in thicker particles represent a significant portion of the area to which binder is applied; however, the binder applied to such surfaces is not believed to contribute materially to the actual bonding of the particles.
(b) The bond between thick particles must also resist greater internal stresses of the board; this usually calls for more adhesive in order to achieve an adequately high bond strength.
(c) Particles with high s values can be spot-welded with relatively sparse adhesive droplets. Thick particles with low s values, on the other hand, require a sufficient amount of adhesive to possess an adhesive layer[6] on their surfaces for adequate particle-particle bonding.

Equation 5–1a gives the slenderness ratio for particles with square or rectangular cross sections. For the case of round particles, i.e., fibers, it will be:

$$s = \frac{l}{d} \qquad (5\text{–}1b)$$

in which d represents the particle diameter. Equation 5–1 notes that a very thin but relatively short particle may have the same s as a very long

5–1. *The influence of particle size on resin take-up when a constant level of strength is considered*

Particle Thickness (mm)	Slenderness Ratio	Particle Surface Area in m² per 100 gm of Oven-dry Particles	Resin Take-up (gm/m²)
0.05	700	9.40	0.85
0.1	350	4.70	1.7
0.3	117	1.57	5.2
0.5	70	0.94	8.6
1.0	35	0.47	17.2

SOURCE: Istrate, V., Filipescu, G. H., and Stefu, C. 1965. The influence of chip dimension on the consumption of glue and on the quality of particleboard.

5–2. Effect of particle slenderness ratio on the modulus of rupture of particleboard when three different levels of resin spray (in grams per square meter of particle surface area) are used.

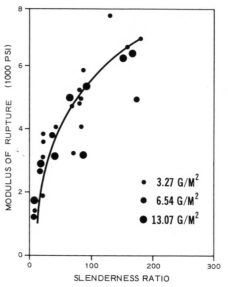

SOURCE: Post, P. W. 1958. The effect of particle geometry and resin content on bending strength of oak particleboard.
Redrawn in modified form by S.I.U. Cartographic Laboratory

and thick one. Also, one particle could have a square cross section while the other—with the same s value—could be very wide. Apart from particle width, s does not account for a number of other particle characteristics. For instance, the bulk density of slivers and flakes with the same slenderness ratios differs markedly, as can easily be demonstrated by freely depositing matchstick-size particles and thin flakes. The mode of travel of these two particle types varies considerably in an air stream. To account, at least in part, for these particle characteristics, another parameter called the flatness ratio, j, is defined:

$$j = \frac{w}{t} \tag{5-2}$$

in which w and t refer to particle width and thickness respectively. For particles with square cross sections, Equations 5–1 and 5–2 yield the same value, that is:

$$j = s.$$

For flat particles (rectangular cross section), we have a situation in which $\dfrac{w}{t} > 1$.

There is not close agreement as to what constitutes optimum value for an industrially practical slenderness ratio. Most of the research results available appear to indicate that, for face particles, slenderness ratio should range from 120 to 200.[7] Particles represented by such ratios are often thin and long, possessing a high degree of pliability, particularly when cut from low- and medium-density species. When a layered construction is used in particleboard manufacture, the slenderness ratio of the particles intended for the core layer is considerably less than those given above, with the optimum being approximately 60.[8]

Figures 5–2 and 5–3 indicate that an increase in the slenderness ratio results in a stiffer and stronger board in bending. Figure 5–3 further

5–3. *Effect of particle slenderness ratio on modulus of rupture, dimensional properties, and water absorption of Douglas-fir particleboard.*

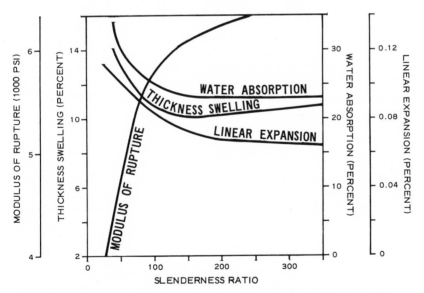

SOURCE: Brumbaugh, J. I. 1960. Effect of flake dimensions on properties of particleboard. Redrawn in modified form by S.I.U. Cartographic Laboratory

5–4. *Effect of particle slenderness ratio
on screw withdrawal load.*

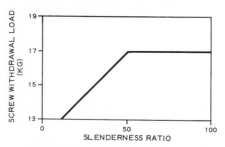

SOURCE: Kimoto, K., Ishimori, E., Sasaki, H., and
Maku, T. 1964. Studies on particleboards. Part 6:
Effects of resin content and particle dimension on
the physical and mechanical properties of low-
density particleboards.
Redrawn in modified form by S.I.U. Cartographic
Laboratory

indicates that increasing the slenderness ratio, up to a limit, also has
desirable effects on water absorption and dimensional stability of the
particleboard. These two figures generally show that, at slenderness
ratios of 150–200, good bending properties along with good board
stability can be obtained. Kimoto and coworkers[9] noted that screw
withdrawal resistance increased with increasing the slenderness ratio up
to a value of 50. Beyond 50, no further increase was observed (fig. 5–4). [10]
However, all available information indicates that increasing slenderness
ratios will result in decreased internal bond strength.[11] The values of
internal bond strength, on the other hand, can be maintained at adequately high levels for all current applications of particleboard at
slenderness ratios required to obtain other desirable values.

5–4. Particle Surface Area

Particle surface area per unit weight is a highly important parameter
which must be considered in resin application if adequate bonding is to
be obtained. Particles possessing greater surface area per unit weight
require a higher amount of resin for sufficient coverage compared to
particles which do not possess such high surface areas for the same unit
weight. The surface area per unit weight of a given particle not only
depends on the density of the wood species from which it is produced,
but also on the particle size. In fact, the latter has an enormous influence
on the surface area per unit weight. For instance a millionfold increase
in surface area results if one reduces a cube measuring one inch, at each

of its dimensions, (total surface area being 6 in.^{2}) into tiny cubes with each side measuring 10^{-6} inch.

The surface area per unit weight can generally be calculated, with some approximation, if the considered particle type possesses a regular geometry. To do this, the basic relationship between weight and volume of any given material is utilized:

$$W = vg$$

with W referring to weight, v to volume, and g to the density in any consistent set of units. Expanding the above equation for, say, a flake we obtain:

$$W = wltg \qquad (5-3)$$

where w, l, and t represent width, length, and thickness of the flake respectively. The flake surface area, a, is calculated as follows:

$$a = 2(tl+wl+tw) \qquad (5-4)$$

which accounts for all of the flake surfaces, including edges and ends. Equations 5–3 and 5–4 are used to obtain the flake surface area per its unit weight, a', as follows:

$$a' = \frac{a}{W} = \frac{2(tl+wl+tw)}{wltg}. \qquad (5-5)$$

In this equation, if all dimensions are considered in centimeters, weights in grams, then a' units will be in cm^2 per gram. In dealing with thin flakes, the first and the last terms inside the parenthesis of equation 5–5 can be dropped due to inconsequential contributions made by these terms. Equation 5–5 will then take the following form:

$$a' = \frac{2}{tg} \qquad (5-6)$$

showing that the surface area per unit weight is inversely proportional to particle thickness and the wood density from which such flakes were generated. This equation confirms that increases in particle thickness and/or density of the wood species will result in a reduction in the surface area per unit weight and vice versa. In equation 5–6, a' is in cm^2 per gram, t is in cm, and g is in gm per cm.^{3} To use the American industrial units, equation 5–6 will be:

$$a'' = \frac{24}{t'g'} \qquad (5-7)$$

in which t' is in inches, g' in pcf, and a'' in ft^2 per pound. Figure 5–5 shows a plot of equation 5–7 for quaking aspen ($g' = 25.1$ pcf at oven-dry condition) which indicates how sharply a'' is affected by particle thickness.

Maloney,[12] working with chunky Douglas-fir particles of mixed sizes, measured surface area per unit weight (one gram) with magnifying glass and microscope, assuming the particles to be rectangular blocks. He segregated his particles according to Tyler screen designations [13] into five sizes (fig. 5–6) and selected 25 representative particles from each designation for actual surface measurements. His data naturally indicated that a substantial difference existed in the surface area per unit weight in the various particle size categories: the finest fraction used (designation: -42) possessed over five times as much surface area per unit weight compared to the coarsest (designation: $-6+9$) fraction. However, Maloney assumed particle surfaces to be perfectly smooth. In our calculations in this section, we also assumed the same. Such an assumption, of course, is not exact for practically any type of wood particle. The actual surface area will be somewhat greater than that accounted for by a' or a''. The extent of discrepancy between these parameters and the actual surface area will be greater as the particle surface becomes rougher. In extreme cases of surface roughness, the actual particle surface area

5–5. *The influence of particle thickness on parameter a″ for quaking aspen.*

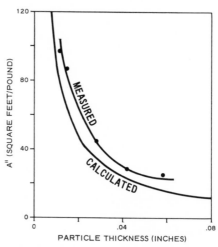

PARTICLE THICKNESS (INCHES)

SOURCE: Gunn, J. M. 1963. Wafer dimension control; number one design criteria for plant producing particleboard for building construction.
Redrawn in modified form by S.I.U. Cartographic Laboratory

5-6. *(Above) Five particle sizes segregated by Tyler screen designations. (Below) Influence of particle size on particle surface area.*

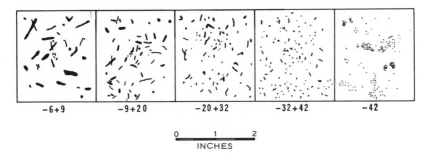

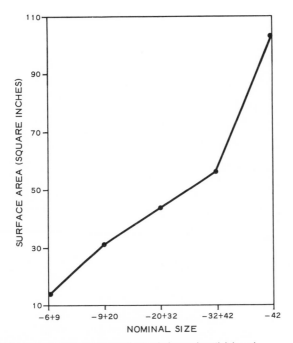

SOURCE: Maloney, T. M. 1970. Resin distribution in layered particleboard.
Redrawn in modified form by S.I.U. Cartographic Laboratory

could conceivably be twice the values calculated or measured under the assumption of perfect smoothness.[14] Thus, the character of particle surface should be carefully examined before trust is placed on calculated values such as a' or a''.

5–2. *The properties of particleboard made of fiberous stock of various sources*

Property	Mixed Hardwoods Medium Density	Mixed Hardwoods Low Density	Oak	Hemlock	Bagasse
Density (pcf)	40.0	25.6	40.3	40.6	41.0
Thickness (in.)	0.800	1.48	0.740	0.830	0.760
Modulus of rupture (psi)	4100	730	2180	3650	3160
Modulus of elasticity (10^3 psi)	360	——	280	310	470
Internal bond (psi)	110	80	110	95	100
Screw holding face—(lbs) edge—(lbs)	385 240	—— ——	230 230	270 240	230 240
Hardness (lbs)	1370	350	1230	1270	1030
Water · · · Thickness Swelling	3.1	——	4.6	4.5	2.3
Soak (24 hrs) Water Absorption	17.5	——	20.6	12.0	15.0
(percent) Linear Expansion	0.18	——	0.17	0.16	0.11

Urea-formaldehyde (8 percent) and 1 percent wax used.

SOURCE: Raddin, H. A. 1967. High frequency pressing and medium density board.

5–5. Other Considerations

During the 1950s, it was believed that, to make a particleboard of acceptable qualities, it must consist of precisely shaped particles. However, a great deal of evolutionary refinement, coupled with a changing economic picture, has made the use of a variety of residues from shavings to sawdust possible and, in some cases, imperative to remain economically competitive. In fact, at the present time, precisely cut flakes make up only a small portion of particle raw material in the United States, with residues, particularly planer shavings and pressure-refined fibers, commanding a substantial lead. The proportion of other types of particles, e.g., splinters and granules, used in particleboard is small. The primary reason for this appears to be that super-smooth panels with adequate strength and other desirable properties can be obtained from shavings and fibers. Table 5–2 shows a set of typical data obtained in manufacturing medium density particleboard using fibers from a number of species and bagasse.

I have noted that per constant amount of resin (based on weight), short, thick particles produce a board of higher internal bond strength compared to one made of long, thin particles.[15] This phenomenon can be explained in part by the higher amount of resin available per unit surface area due to lower surface area per unit weight. Brumbaugh[16] also postulates that a greater degree of discontinuity in horizontal planes in the board, in such cases, contributes to higher strength values. He states that the more continuous horizontal planes create zones of weakness, resulting in lower strength. Short, thick particles normally used in the core layer require a greater amount of resin per unit surface area as compared to long, thin particles or very fine granular particles used for surface layers. Board surfaces are usually of higher density compared to the core. This provides for a more efficient resin utilization because of intimate interparticle contact. Longer and thinner particles, everything else being the same, produce a board with high bending strength and dimensional stability. Shorter and thicker particles produce a board with lower bending strength and dimensional stability but higher internal bond strength. The layered structure for particleboard was engineered to take advantage of the opportunity provided by this phenomenon and produce a board with both high bending and internal bond strengths. Layering is accomplished by placing the type of particles that generate higher bending strength on the surface layers where such stresses are at their maximum while placing, in the core, the type of particles capable of producing high resistance to internal bond stresses. Board structure of this sort also makes economic sense in production since thicker, cruder particles normally costing less are hidden within the board structure and will not detract from surface appearance.

Obviously, the type of raw material received by a given plant has relevance to the type of board structure which can be produced. From the standpoint of a board's bending strength, the least (first) to most desirable forms of raw material can generally be listed as follows:

(a) Sawdust, planer, and sanderdust.
(b) Hammermilled, dry planer shavings.
(c) Hammermilled, moist planer shavings.
(d) Chipped and hammermilled dry solid wood.
(e) Chipped and hammermilled undried solid wood.

(f) Fiber clusters and fibers.

(g) Dry solid wood cut to flakes.

(h) Undried solid wood cut to flakes.

Bending strength, of course, is not the only criteria that must be taken into consideration in particleboard production and, therefore, the above list of raw material desirability may have to be reshuffled by a given plant.

Notes

1. Brumbaugh, J. I. 1960. Effect of flake dimensions on properties of particleboard.
   Heebink, B. G., and Hann, R. A. 1959. Stability and strength of oak particleboards.
   Johnson, E. S., ed. 1956. Wood particleboard handbook.
   Marra, G. G. 1954. Discussion of article by Turner.
   Turner, D. H. 1954. Effect of particle size and shape on strength and dimensional stability of resin-bonded wood particle panels.
2. Some types of thick flakes are referred to as wafers.
3. Heebink, B. G., Hann, R. A., and Haskell, H. H. 1964. Particleboard quality as affected by planer shaving geometry.
4. Ibid.
5. Istrate, V., Filipescu, G. H., and Stefu, C. 1965. The influence of chip dimension on the consumption of glue and on the quality of particleboard.
6. Ibid.
7. Brumbaugh, J. I. 1960. Effect of flake dimensions on properties of particleboard.
   Istrate, V., Filipescu, G. H., and Stefu, C. 1965. The influence of chip dimension on the consumption of glue and on the quality of particleboard.
   Post, P. W. 1961. Relationship of flake size and resin content to mechanical and dimensional properties of flakeboard.
8. Istrate, V., Filipescu, G. H., and Stefu, C. 1965. The influence of chip dimension on the consumption of glue and on the quality of particleboard.
9. Kimoto, K., Ishimori, E., Sasaki, H., and Maku, T. 1964. Studies on particleboards. Part 6: Effects of resin content and particle dimension on the physical and mechanical properties of low-density particleboards.
10. Luan flakes and Japanese industrial standards were used in manufacture and testing.
11. Brumbaugh, J. I. 1960. Effect of flake dimensions on properties of particleboard.
    Kimoto, K., Ishimori, E., Sasaki, H., and Maku, T. 1964. Studies on particleboards. Part 6: Effects of resin content and particle dimension on the physical and mechanical properties of low-density particleboards.
    Post, P. W. 1958. The effect of particle geometry and resin content on bending strength of oak flakeboard.
12. Maloney, T. M. 1970. Resin distribution in layered particleboard.
13. In such designations, $-9 +20$ for instance means that the particles pass through mesh 9 but are held by mesh 20.
14. Post, P. W. 1961. Relationship of flake size and resin content to mechanical and dimensional properties of flakeboard.

15. Berchem, T. E. 1970. The effect of particle size on strength, dimensional properties and surface roughness of hickory particleboard.

Brumbaugh, J. I. 1960. Effect of flake dimensions on properties of particleboard.

Halligan, A. F. 1969. Recent glues and gluing research applied to particleboard.

Lehmann, W. F. 1965. Improved particleboard through better resin efficiency.

Plath, E. 1963. Influence of density on the properties of woodbase materials.

Talbott, J. W., and Maloney, T. M. 1957. Effect of several production variables on the modulus of rupture and internal bond strength of boards made of green Douglas-fir planer shavings.

16. Brumbaugh, J. I. 1960. Effect of flake dimensions on properties of particleboard.

Binders

6-1. Urea-Formaldehyde Adhesives

Due primarily to their low cost, versatility, and ease of application, urea-formaldehyde adhesives are the most widely used resins in the particleboard industry. The properties offered by these resins are adequate for bonding particles in products intended for interior applications (e.g., in subfloor and furniture core stock). Many improvements along with a continuous reduction in their cost, have kept urea-formaldehyde adhesives in a leading position during the past several decades. Further research in producing yet better properties will likely help urea resins retain their market position in the field of wood particleboard manufacture.

Urea itself is the most prominent member of the amino resins with the chemical structure:

$$H - N - H$$
$$|$$
$$C = O$$
$$|$$
$$H - N - H$$

Urea adhesives are condensates of urea and formaldehyde HC HO (formalin) which have been advanced in polymerization to a degree that still allows almost complete dispersion in water at solid contents of about

50–65 percent. Mole ratios of approximately 1.5 to 2 of formaldehyde to one of urea are common for urea-formaldehyde resins which can be produced in dispersion or spray-dried forms.

To produce urea, ammonia forms a basic raw material. When ammonia reacts with carbon dioxide under pressure, an intermediary product called ammonium carbamate is produced, generating urea upon heating at 135°C.:

$$2NH_3 + CO_2 \rightarrow NH_4\ CO_2\ NH_2 \rightarrow H_2N - \overset{\displaystyle O}{\overset{\displaystyle \|}{C}} - NH_2 + H_2O$$

ammonium carbamate   urea

The other component needed to produce the desired adhesives, namely formaldehyde, is produced from methyl alcohol (CH_3OH) also known as methanol. Although methanol can be produced in a number of ways, an example follows:

$$C + H_2O \rightarrow CO + H_2$$

$$CO + 2H_2 \leftrightarrows CH_3OH$$

$$CH_3OH \leftrightarrows CH_2O + H_2$$

The dehydrogenation of methanol to formaldehyde is accomplished in the presence of copper, silver, or platinum catalyst. The presence of air burns H_2 to H_2O, supplying necessary heat for the operation. In the production of methanol shown above, temperatures of 300–400°C. in the presence of catalysts are used. To forestall the reverse reaction in methanol production, pressures of 100 atmospheres and above and temperatures of 350°C. or greater are often employed. The formaldehyde produced in the manner described is in gas form under normal temperatures and is usually dissolved in water to form a solution referred to as formalin.

Formaldehyde can also be produced from the methane (CH_4) contained in natural gas. Methane is catalytically oxidized usually at 500–700°C. in the presence of oxides of nitrogen catalysts:

$$CH_4 + O_2 \rightarrow HCHO + H_2O$$

At the high temperatures used in producing formaldehyde from methane, the former is readily decomposed and since the reaction is exothermic, very careful control in the production must be exercised.

The simplest reaction product of urea and formaldehyde is brought about when equimolar proportions of urea and formaldehyde are used. This reaction produces a type of "building block" for urea adhesives called monomethylol urea:

$$
\begin{array}{ccc}
\text{NH}_2 & & \overset{\displaystyle \text{H}}{\underset{|}{\text{N}}} - \text{CH}_2\text{OH} \\
| & & | \\
\text{C} = \text{O} + \text{HCHO} & \rightarrow & \text{C} = \text{O} \\
| & & | \\
\text{NH}_2 & & \text{NH}_2
\end{array}
$$

monomethylol urea

Under proper conditions, the reaction results in the attachment of the formaldehyde to a nitrogen of the urea molecule. If the molar ratio of formaldehyde to urea is increased to 2:1, then dimethylol urea, another type of "building block" for urea-formaldehyde adhesives, will result:

$$
\begin{array}{ccc}
\text{NH}_2 & & \overset{\displaystyle \text{H}}{\underset{|}{\text{N}}} - \text{CH}_2\text{OH} \\
| & & | \\
\text{C} = \text{O} + 2\text{HCHO} & \rightarrow & \text{C} = \text{O} \\
| & & | \\
\text{NH}_2 & & \underset{\displaystyle \text{H}}{\overset{|}{\text{N}}} - \text{CH}_2\text{OH}
\end{array}
$$

dimethylol urea

Formaldehyde is again attached to the nitrogens of the urea molecule. Both monomethylol urea and dimethylol urea can be produced in aqueous solutions using an alkaline catalyst.

Polymerization (curing) of the monomeric "building blocks," described above, is believed to involve different and complex mechanisms depending upon the reaction conditions and the molar ratio of formaldehyde and urea used. Analysis of intermediate condensates in the polymerization process of urea-formaldehyde resins indicates the presence of methylene and ether bridges between the urea molecules:

$$
\cdots - \overset{\displaystyle \text{H}}{\underset{\displaystyle \underset{|}{\text{C}=\text{O}}}{\underset{|}{\text{N}}}} \rule{1.5em}{0.4pt} \overset{\displaystyle \text{H}}{\underset{\displaystyle \text{H}}{\underset{|}{\text{C}}}} \rule{1.5em}{0.4pt} \overset{\displaystyle \text{H}}{\underset{\displaystyle \underset{|}{\text{C}=\text{O}}}{\underset{|}{\text{N}}}} \rule{1.5em}{0.4pt} \overset{\displaystyle \text{H}}{\underset{\displaystyle \text{H}}{\underset{|}{\text{C}}}} - \cdots
$$

$$
\begin{array}{cccc}
| & & | & \\
\text{NH}_2 & & \text{NH}_2 &
\end{array}
$$

methylene bridge

$$
\begin{array}{ccccccc}
H & & H & & & H & & H \\
| & & | & & & | & & | \\
\cdots -N & - & C & - O - & C & - N & - \cdots \\
| & & | & & & | & & | \\
C=O & & H & & & H & & C=O \\
| & & & & & & & | \\
NH_2 & & & & & & & NH_2
\end{array}
$$

ether bridge

The polymerization may, for instance, advance through the formation of an intermediate compound called methylene urea

$$
\begin{array}{l}
N = CH_2 \\
| \\
C = O \\
| \\
NH_2
\end{array}
$$

in the presence of dilute acids which will subsequently condense to generate a linear polymer utilizing one of the bridges indicated above. Terminal groups may exist as amide ($-NH_2$), methylol ($-CH_2OH$) or azomethine ($-N = CH_2$). Direct determination of the molecular size in cured urea resins points to a relatively low degree of polymerization (compared to, for instance, thermoplastic adhesives). During transition from monomeric to polymeric state whether under the influence of heat,[1] catalyst, or both, the urea resins split out water of condensation. Urea resins thus belong to those classed as condensation polymers in contrast to addition polymers such as polyesters which do not eliminate water as they advance from the monomeric to the polymeric state.

The urea resins manufactured for the particleboard industry are either essentially monomeric or, at most, only slightly advanced in polymerization. The low degree of polymerization is necessary to provide solubility of the resin product in the initial stages of application. Available urea-formaldehyde resins contain the reactive terminal groups enabling them to condense further in an acid environment under the influence of heat. Urea resins are now tailored to meet the needs of individual processes and wood raw materials (see section 6–5). The resin supplied to the industry is usually employed with little or no dilution and is catalyzed to yield short press cycles. Urea-formaldehyde is white to nearly colorless, thus imparting no color to the resulting panel. However, urea-bonded boards are not durable in wet or damp environments and break down under heat, a characteristics which has severely limited their use for particleboard intended for exterior applications. All components of the resin system in the manufacture of urea-formaldehyde are produced with relatively uniform quality. Changes occurring in the resin during transport and storage are not normally critical if a reasonable amount of care is exercised (see section 6–5).

Reaction kettles equipped with turbo-agitators and reflux condensers are used to manufacture urea-formaldehyde resins. Such kettles are also jacketed for heating and cooling. A variety of products are made by changing, for example, the mole ratio of the reactants, pH and buffering adjustments during reaction or the degree of polymerization obtained in the condensation cycle. Selecting the mole ratio of formaldehyde to urea desired, the needed amount of urea is entered into the kettle containing formalin. The added urea, which may be charged to the reactor at the start of the reaction or stepwise, will dissolve in formalin easily and endothermically. Sodium hydroxide or some other alkali is utilized to control the pH of the mixture to about 7.5 to 8.0, which is subsequently heated to about 98° C. The reaction is allowed to continue until the desired viscosity (a measure of the degree of resin polymerization) is obtained. The rate of polymer growth is increased by lowering the reaction pH to 5.6–6.0 (using formic acid). Once the desired level of viscosity is obtained, the solution pH is readjusted to about 8.0 (to forestall further polymer advance) and cooled. The resin is now ready for shipment by tank truck, tank railroad car, or drums. Urea-formaldehyde resins made in the preceding manner will have a solids content ranging from 45 to about 65 percent.

Polymerization of urea-formaldehyde can occur with or without heat in an acid environment and is inhibited in alkaline media. Although molecular weight distribution and chemical homogeneity are perhaps the most important factors contributing to the individuality of urea-formaldehyde adhesives, others are: amount of free formaldehyde, buffer compounds and their effects, water tolerance, manufacturing methods, and molar ratio of formaldehyde to urea. Curing agents, also called hardeners, accelerators, or catalysts are chemical substances added to the urea-formaldehyde resins usually by the user and just prior to application to speed up polymerization. These chemicals are either acidic substances by themselves or can liberate acids when mixed with the adhesive. In particle-board manufacture, heat is used in addition to the catalyst, to increase the rate of cure. The most widely used catalysts are ammonium salts of strong acids because they are inexpensive, convenient to handle, and yield a high ratio of working life to cure time. Ammonium salts used for this purpose, usually include the salts of chloride, sulfate, phosphate, nitrate, fluoride, and borate. In addition, ammonium thiocyanate can be an efficient hardener but is not used because it is poisonous. Also, alkanolamine salts (aliphatic-substituted hydroxylamine salts such as 2-amino-2-methyl propanol-1 hydrochloride) may be used as hardeners.

Alkanolamine salts are considered general-purpose hardeners for urea and other amino resins. They are believed to be an improvement over the ammonium salts since they release their acidic constituent more slowly, thus providing a more gradual resin cure and thereby reducing the chance of precure. There are also inorganic, nonammoniacal salts (such as magnesium chloride) that are effective as hardeners. Furthermore, weak organic acids are employable for this purpose but have had only limited use.

The commonly used ammonium salts function as hardeners by reacting with any uncombined formaldehyde (free and/or liberated under curing conditions) present in the resin as well as with the terminal methylol groups of the monomeric and low polymeric forms to free the corresponding acid, hexamethylenetetramine (hexamine), and water. The structure of hexamethylenetetramine, on the basis of X-ray studies, is generally credited with a cyclic form as follows:

$$
\begin{array}{c}
\text{N} \\
\diagup\;|\;\diagdown \\
\text{CH}_2 \quad \text{CH}_2 \quad \text{CH}_2 \\
|\qquad\quad|\qquad\quad| \\
\qquad\;\text{N} \\
\diagup\;\diagdown \\
\text{CH}_2 \quad \text{CH}_2 \\
\diagup\qquad\qquad\diagdown \\
\text{N}\qquad\qquad\qquad\text{N} \\
\diagdown\qquad\qquad\diagup \\
\text{CH}_2
\end{array}
$$

For an example, let us consider a case in which ammonium chloride is utilized as a hardener. The reaction with formaldehyde in this particular instance will release hydrochloric acid, hexamethylenetetramine and water as shown below:

$$4\,NH_4\,Cl + 6\,HCHO \rightarrow 4\,HCl + (CH_2)_6\,N_4 + 6\,H_2O$$

The liberation of acid in this reaction results in an immediate decrease of pH value of the adhesive. Additionally, the liberation of formaldehyde from methylol groups causes further drop in the pH value. Ammonium salts are good hot hardeners because formaldehyde liberation is sharply increased as the urea-formaldehyde adhesives containing such hardeners are heated. Figure 6–1 clearly shows this phenomenon: the ammonium chloride is added and the urea resin temperature is increased from 21°C. to 40°C., resulting in a sharp drop in pH.

To extend the working life of urea-formaldehyde adhesives, it is customary to add a pH buffer such as ammonium hydroxide to slow down the rate of polymerization at room temperature. Buffers are chemical

6–1. *Decreasing urea resin pH upon adding ammonium chloride and maintaining the mixture at three different temperatures.*

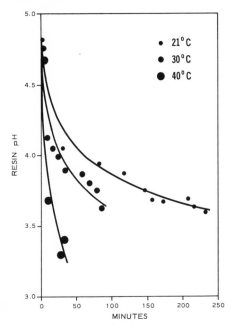

SOURCE: Rayner, C. A. A. 1965. Synthetic organic adhesives.
Redrawn in modified form by S.I.U. Cartographic Laboratory

substances often added to urea resins to increase their resistance to pH change as these resins come in contact with acidic or basic materials. Improper or excessive buffering, however, can be disadvantageous since drops in the adhesive pH will be resisted by the buffer, resulting in a slow cure. Buffering, when correctly done, increases the adhesive working life and serves to prevent the pH of the adhesive in the particle-particle bond from dropping to highly acidic values, which can degrade the resulting particleboard.

Buffering of urea adhesives can be made to occur at various pH levels, as examples show in figure 6–2. This figure sets the centering of buffering levels for the three curves at pH values of 3.5, 5.5, and 7.5. It shows that the urea-formaldehyde resins under consideration are heavily buffered since a large quantity of standard acid or base is required to change measurably their pH value. In the examples shown in figure 6–2, the urea

6–2. *Examples of urea resin pH drops when buffered at average pH levels of (A) 7.5, (B) 5.5, and (C) 3.5.*

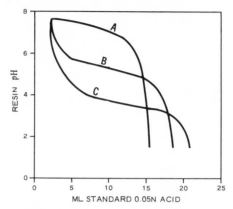

SOURCE: Lambuth, A. L. 1967. Performance characteristics of particleboard resin binders. Redrawn in modified form by S.I.U. Cartographic Laboratory

resin buffered at the pH value of 3.5 will rapidly cure at lower temperatures as it is exposed to the natural acidity of wood species. However, the danger of considerable advance in resin polymerization is present from the time the resin is shipped to the time when the mat is under full pressure in the hot press. In contrast, when the resin is buffered to have a central pH value of 7.5, it is capable of maintaining an extended storage life without the risk of premature polymerization. To get fast curing rate in the hot press, the natural acidity of wood alone may not be sufficient. Addition of hardeners to the resin is needed. For the urea-formaldehyde adhesive buffered to have a central pH value of 5.5, we can, therefore, expect a state somewhere in between central pH values of 3.5 and 7.5. It is customary to buffer urea-formaldehyde adhesives so that their pH values remain near neutral when the resin cannot be consumed rapidly and when storage stability is a prime requisite.

Powdered urea resins may contain an appropriate catalyst when supplied. Similar to the liquid form, they are cured under the application of heat. Powdered resins are usually capable of providing a degree of flow in the hot press before they are hardened. Spray-dried resins can cure at the same fast rates as the liquid form in the hot press. Although it is usually an expensive procedure, powdered urea-formaldehyde resin is sometimes used to raise the solids content of corresponding liquid resin.

6-2. Phenol-Formaldehyde Adhesives

Phenol-formaldehyde adhesives are the second most widely used binders in the production of particleboard. They are not nitrogen-containing and have some basic performance differences with urea-formaldehyde resins. Phenolic resins, unlike urea-formaldehyde, are durable, exhibiting resistance to breakdown in both cold and hot water. Phenolics are not attacked by weak acids or alkali and are resistant to common organic solvents. These adhesives also show good resistance to higher temperatures. However, phenolic-formaldehyde adhesives are at a disadvantage compared to urea binders in a number of respects including solids content, storage life, cure speed, tack characteristics, compatibility with additives including wax emulsions, and cost. Furthermore, they can impart color to the resulting board.

Continuous research and development to improve the properties of phenolic resins has paralleled that of ureas. The result has brought about improvements in such important properties as cure rate and cost. This does not mean, however, that phenolics have overtaken the ureas: they still require somewhat longer press times, higher curing temperatures, and cost more. Thus, the volume of phenol-formaldehyde resins used by the particleboard industry remains far behind that of urea-formaldehyde.

The two chemical components required for the manufacture of phenolic adhesives are phenol and formaldehyde. Phenol can be obtained from such sources as coal-tar which is distilled into three fractions, one of which (a creosote oil fraction) contains phenol. Also, phenol can be commercially synthesized in various ways. For instance, an older process involves the sulfonation of benzene and hydrolysis by fusion with sodium hydroxide:

In another process, benzene can be chlorinated to chlorobenzene with further hydrolysis by heating with sodium hydroxide in the presence of a copper catalyst:

In yet another process, benzene can be oxidized with air in the presence of hydrochloric acid to chlorobenzene and eventually to phenol as follows:

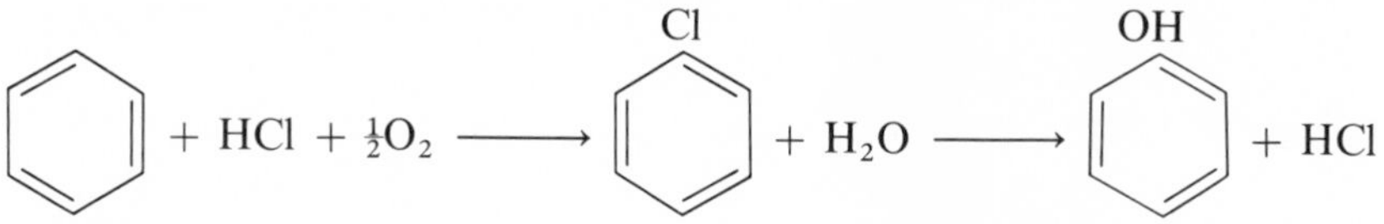

As is noted from the above reaction, the hydrochloric acid is regenerated. There also are other means of phenol production.[2]

Phenol-formaldehyde adhesives are condensation products of formaldehyde and phenol involving highly complex organic chemical processes. There are many ways for phenol and formaldehyde to combine in this condensation process, with products being either thermosetting or thermoplastic. The manner in which phenol and formaldehyde can combine depends on many factors such as type and amount of catalyst, the reaction pH, the molar ratio of phenol to formaldehyde, the time and temperature of the reaction, and the use of various additives. The thermosetting products of phenol and formaldehyde reaction are formed when formaldehyde is used in molar excess in an alkaline field.

The first reaction between phenol and formaldehyde can generate products such as ortho- or para-monomethylol phenol:

orthomonomethylol phenol

paramonomethylol phenol

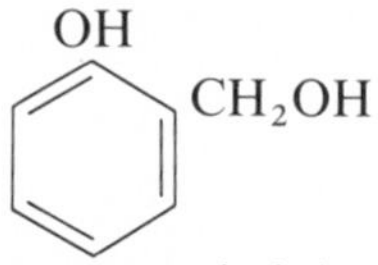

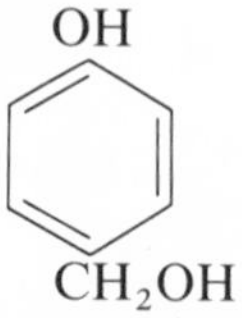

These can subsequently react with more formaldehyde to yield di- and tri-methylol phenols:

2, 4-dimethylol phenol

2, 6-dimethylol phenol

2, 4, 6-trimethylol phenol

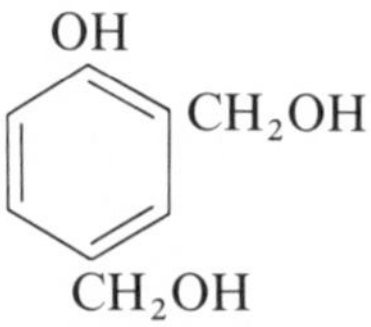

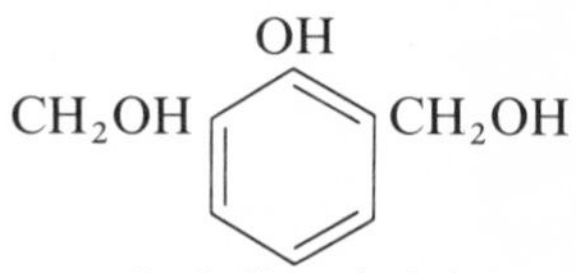

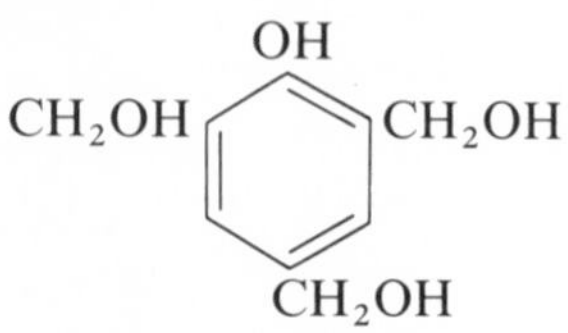

In the ensuing reaction, a methylol group of one molecule reacts with a second phenolic molecule, splitting off water, to form a dihydroxydi-phenylmethane type of compound as those shown below:

Or, it can react with a methylol group attached to another phenolic molecule, splitting off water and producing a dihydroxydibenzylether such as:

In either alkaline or acid environments this ether splits off formaldehyde to form a methylene-linked compound, with the formaldehyde becoming available for further reaction.

In the production of thermosetting phenolic resins, the molar ratio of formaldehyde to phenol usually ranges from 1:1 to 3:1. Catalysts employed for these resins are usually strong alkali (e.g., sodium hydroxide) although ammonia is also frequently used. Since the reaction between phenol and formaldehyde involves condensation polymerization, as pointed out earlier, the weight of the cured resin will be less than the combined weights of reactants by the amount of water split off. This amount of water released is proportional to the extent of polymerization and number of crosslinks. Condensation proceeds slowly at room temperature and rapidly at elevated temperatures.

Phenolic resins are comparatively brittle in the cured state. This characteristic is not significant in particleboard production where thick glue lines are not generally encountered. The color generated by phenolics is considered a drawback. In recent years the addition of white pigments has helped to alleviate this problem to some extent. Phenol-formaldehyde adhesives are tolerant to such additives as fungicides, water repellents, and fire retardants. Phenolic resins used by the industry have a high pH value when stored and used. The curing speed depends to a large degree on their pH: high alkalinity accelerates polymerization.

Phenolic resins, similar to the ureas, can be manipulated in many ways to meet the requirements of a given operation. For instance, by changing the reactant ratios and the catalyst level, a large variety of resins are produced. Once other variables, i.e., time, temperature, and the type and amount of additives are considered, much can be done to meet the needs of particleboard production. Phenol-formaldehyde adhesives can be obtained in both liquid and powder form. The former has a solids content of up to 40 percent. Some have attributed the efficiency of phenolic resins to be due partly to their greater degree of dilution, which allows good distribution during blending.[3] In some cases in commercial board manufacture, it has been observed that the board strength with 6 percent urea-formaldehyde adhesive (based on oven-dry weight of particles and resin solids) is equated with one containing only 4.5 percent phenol-formaldehyde, with other factors remaining constant.

The powder form of phenolic resins may contain an appropriate catalyst and is capable of melting and flowing in the hot press before it is cured. Powdered phenolics are utilized to a large extent by the molding industry. As in the case of urea resins, powdered phenolics can also be used to increase the solids content of a liquid resin of its corresponding type. Cross mixtures of urea and phenolic resins are not practical since each interferes with the curing mechanism of the other. Phenolic-bonded particleboard is being made on an increasing scale. In the production of exterior particleboard, phenolics are almost always used. In Europe, considerable impetus has been given to the production of phenolic-bonded boards by the new German regulations in effect since January 1, 1970, requiring that boards used for structural load carrying purposes be bonded with phenol-formaldehyde resins.[4]

6–3. Melamine-Formaldehyde Adhesives

Melamine-formaldehyde is the most durable and most expensive of the amino resin class. It is not usually used by itself in the manufacture of wood particleboard. Since it offers better moisture and heat resistance than the urea-formaldehyde resins, it is sometimes used in conjunction with ureas to upgrade the binder. This improvement in moisture and heat resistance, however, is not adequate for exposing urea-melamine-formaldehyde-bonded boards to the environment of the outdoors: both urea- and urea-melamine-bonded particleboards disintegrate when subjected to the ASTM Accelerated Aging Exposure D1037.[5] Nevertheless, the slight improvements in the board properties, compared to the straight urea-bonded boards, make the panels qualified for more severe interior applications. The mixing of urea and melamine resins was more prevalent in earlier years of wood particleboard manufacture.

The basic raw materials required for the manufacture of melamine-formaldehyde are melamine and formaldehyde. Melamine (2, 4, 6-triamino-1, 3, 5-triazine) entered the field of commercial production in

1939 in the United States. It is a white crystalline chemical which can be produced from calcium cyanamide. The latter is, in turn, made from calcium carbide and nitrogen. The cyanamide, which is liberated in the presence of acids, is polymerized to produce dicyandiamide upon heating with further polymerization, resulting in the formation of melamine:

dicyandiamide   melamine

The formation of melamine in the above process is usually carried out under elevated temperatures (100–400°C.) and pressures (10–100 atmospheres) in the presence of anhydrous ammonia. However, heating dicyandiamide (300°C.) at normal atmospheric pressures also forms melamine.

The reaction between melamine and formaldehyde is believed to leave the triazine ring structure in melamine intact: the resinification takes place by the condensation of one or more of the amino side groups with formaldehyde. The reaction products of melamine and formaldehyde involve the generation of a variety of methylol melamines, depending on the mole ratio of the reactants. For instance, when one mole of melamine is reacted with 3 moles of warm formalin until a solution is obtained, followed by quick cooling, crystalline trimethylol melamine can be obtained:

Trimethylol melamine

Methylol melamines are capable of condensing to form resinous polymers by heating under mild conditions. The presence of small amounts of acid will accelerate the heat cure. Similar to the ureas, the degree of poly-merization prior to the application is low in melamine resins. Melamine-urea resins are hot-pressed to obtain quick polymerization. A bond formed by using melamine-urea-formaldehyde is characterized by im-proved boil, heat, and chemical resistance (compared to one formed by urea-formaldehyde) and a neutral pH.

6–4. Other Adhesive Binders

Among other organic binders which have held some promise for use in the wood particleboard industry, sulfite waste liquor and bark extracts are noteworthy. As of early 1970s, two West European mills were using sulfite waste liquor in their process with some success. This liquor is a basically ligneous by-product of the sulfite process of pulp manufacture and is highly complex in chemistry. Sulfite waste liquor binder has a solids content of about 50 percent with some catalyst added before it is blended with the wood particles. The most effective catalyst so far found is citric acid, although small amounts of urea-formaldehyde can be employed instead.

The sulfite waste liquor is sprayed on the wood furnish in a similar manner as other bonding agents. European mills using this binder incorporate it at the rate of about 12 percent (based on the solids content and the weight of the oven-dry furnish). Wax sizing and fire retardants such as borax is incorporated in a similar fashion as with commonly used adhesives. Heat and pressure in the hot press accomplish resin poly-merization with a long press time, approximately some five times that required for urea resins.[6] After the boards emerge from the hot press, they must be autoclaved for a period of approximately sixty to ninety minutes at 390°F. under about ten atmospheric pressures to complete the reaction. Modifications of this after-press treatment, e.g., autoclave time, influence the board color and other properties. Boards with good strength and improved dimensional stability (compared to those bonded with urea resins) are made with sulfite waste liquor.[7] Waste-liquor-bonded boards are said to be stable under exterior exposure conditions. No apparent surface deterioration is discernable, although some change in surface color and staining around screws and fittings occur after several years of exposure. The boards are easily paintable with oil-based paints.

Some problems associated with after-press treatment have been encountered. These include the corrosion of valves and pipes, pipe block-age by deposits, and temperature stratification in the autoclave (held responsible for the irreversible bowing of large panels). Also, the release of acidic malodorous gases, when the autoclave is opened, generates air pollution problems. Trapping these acidic gases in water has not proved successful. However, the use of alkali solutions has yielded good results. Nevertheless, technical problems, along with market acceptability, must be worked out economically if the sulfite-liquor-bonded boards are to become more prevalent. The low cost and availability of sulfite waste liquor is also a prime consideration in making this process viable. The advantages of being able to utilize waste in obtaining a durable and easy-to-machine panel gives the process some merit.

Interest has also been expressed on the use of bark extracts as binders for the particleboard industry. Bark extracts from various species, e.g., radiata pine,[8] ponderosa pine, tanoak,[9] and wattle,[10] among others, possess a significant quantity of complex, tanninlike constituents which are reactive to formaldehyde and have potential value in the manufacture of resins. Aqueous bark extract solutions, even at low concentrations (as little as 15 percent extract), can react with 1–2 percent formaldehyde (based on extract weight) and can become solid gels upon standing at room temperature. Bark extracts are believed to be a mixture of similar polymers, perhaps differing only in the degree of polymerization. Phenolic functional groups are believed to be present in bark extracts including the following flavonoid structural unit:

Bark extract-formaldehyde adhesives have serious drawbacks compared to commonly used synthetic resins: high viscosity and a very short working life make the solids content of over 30 percent impractical for use in the particleboard industry. Much research is needed to make bark extract economically and technically viable.

6–5. Resin Properties and Requirements

As the wood particleboard industry has grown over the last three decades, its products have improved substantially in quality. Availability of better synthetic resins has largely been responsible for this development.

Urea and phenolic resins have been able to meet many requirements imposed by the manufactures. The need for custom formulating resins based on a large number of factors is now well established. Some of the more important factors are as follows:

1. *Viscosity:* If the viscosity is not in the proper range, it can hamper efficient operation of the resin application system resulting in poor resin distribution in the blending operation. Automatic metering devices and dispensers are particularly vulnerable. Resin viscosity is affected by a number of variables including temperature. Increased ambient temperatures result in lower viscosities for both urea and phenolic resins (fig. 6–3). Tests on resin viscosity, under appropriate conditions, can reveal information pertinent to its age, quality, and, in many cases, wetting characteristics. Tests can further reveal changes in density, stability, solids content, and molecular weight. The determination of viscosity involves

6–3. *Interrelationships between resin solids, resin temperature and viscosity.*

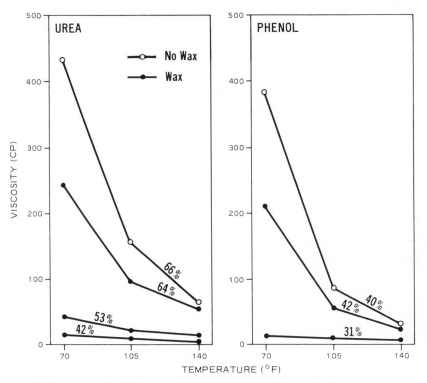

SOURCE: Lehmann, W. 1965. Improved particleboard through better resin efficiency. Redrawn in modified form by S.I.U. Cartographic Laboratory

a simple test using a viscometer generally composed of a stand, a mounted electric motor, and a rotating shaft to which a spindle is attached. The results are registered in centipoise (cp) by an instrument incorporated into the viscometer (see ASTM D1084–60 Method B and ASTM D1286–57). It is generally recommended that the resin viscosity be in the 100 to 500 cp at 70°F. for particleboard production.

2. *Solids content:* Urea- and phenol-formaldehyde resins contain a certain percentage of solids dispersed in a water medium. In particleboard production, the solids content should be high enough such that upon completion of the blending operation, excessive amounts of water have not been added to the wood furnish. The solids content on the other hand can not be excessively high since the resin becomes too viscous, causing problems associated with high resin viscosity already mentioned. The solids content of resins supplied to the particleboard industry is about 65 percent for urea-formaldehyde and 40–50 percent for phenolic resins. Solids content may be determined by weighing a small amount of the adhesive in a clean container and heating until a constant cured weight is obtained. The sample is reweighed and the solids content is calculated as follows:

$$S = \frac{100\ w'}{w}$$

in which S represents the solids content in percent, w and w' are the weights before and after the resin cure respectively. In solids content determinations, the data normally include the cure temperature, oven type (gravity or forced air), number of samples taken, sample size, dish diameter, solids content of each sample, and the mean of the samples of a given run. This determination also yields information on the amount of volatiles released during the hot pressing operation.

3. *Tack:* This property refers to the ability of the adhesive to develop a certain amount of adherence on contact. With the development of caulless production lines and other modifications in the manufacture of particleboard, proper tack development has become a critical requirement (see section 6–6).

4. *Cure:* The cure rate of the resin obviously influences the rate of production. Much research and development has been devoted to reducing the time required in the hot press. Sharp reductions in press time have been achieved. The use of slow curing resins leads to an uneconomical underutilization of the hot press and thereby of the rest of the equipment in the particleboard plant. However, the rate of cure of a given resin must

not be so fast as to produce precure before the full pressure is applied to the mats. Further, good resins will not stick to the cauls or platens once cured.

5. *Formaldehyde release:* The odor of formaldehyde is repugnant and, when excessive, is considered a serious nuisance. All urea- and phenol-formaldehyde adhesives release varying amounts of formaldehyde during hot pressing and later from the products utilizing such binders (see section 6–7).

6. *Storage life:* In particleboard production, adhesives are stored in quantity and eventually used over a period of time. During this period, it is imperative that the resin properties essential for quality board production not be adversely altered. The resin must have a storage integrity such that no rapid viscosity buildup occurs. In addition, its other essential properties, e.g., cure speed and tack characteristics, must not change substantially from the time it is manufactured to the time it is used

6–4. *Effect of storage temperature on storage life of a urea resin.*

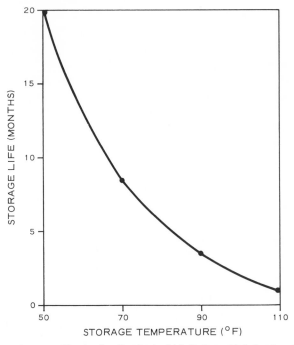

Courtesy of Borden, Inc. Drawing by S.I.U. Cartographic Laboratory

by the board plant. The temperature of the environment in which the resin is stored has a substantial influence on its storage life. Commercial resins have a storage life of a number of months if kept under relatively cool temperatures. Figure 6–4[11] illustrates an example on the influence of temperature upon storage life of a urea-formaldehyde resin indicating a life of some 20 months when the ambient temperature is maintained at around 50°F.[12] During hot summer months, to prevent change during transportation, the resin is usually precooled before shipping in insulated tanks. Should the storage time be long (low resin usage or infrequent resin delivery), adhesive storage tanks are insulated to maintain low resin temperature. In cases where high ambient temperature is coupled with slow resin usage, serious resin degradation can occur unless a cooling system is installed. Apart from ambient temperature in the resin storage area, other factors such as the solids content, resin pH, the degree of buffering, the mole ratio of reactants, and the degree of reaction affect the storage life of resins. For instance, it is possible to produce a resin of long storage stability by using a higher molar ratio of formaldehyde at the cost of a greater degree of formaldehyde release. Also, there is a limit in obtaining a better storage life by buffering since this will be ineffective beyond the point needed to counteract the acid formed by oxidation of formaldehyde. It must be pointed out that it is helpful to buffer a urea resin to help maintain its pH value near neutral (7 to 8). Buffering, while useful in this respect, can become a hinderance to quick resin polymerization after the resin is applied. The pH value alone does not reveal all the information needed to judge the storage life of a resin. For instance, a highly reacted, short-life adhesive can deteriorate seriously during storage without showing any marked change in its pH value. Furthermore, the mechanism of resin failure may not be due to resin advancement but an alkaline rearrangement called *pasting*.[13]

7. *Working life:* This property is the indicator of time which can lapse between the moment the adhesive is ready for use and the time when it is no longer suitable for use. A number of factors, i.e., adhesive age, the temperature encountered by the resin during transportation, storage, and application, and the precision with which a resin is catalyzed, influence the working life of an adhesive. The working life begins in particleboard manufacture when catalysts and additives are added until the particle mats are being consolidated under full pressure and heated in the hot press. The resin suppliers and manufacturers, based largely on experience, usually provide the board plant with information on the length of the working life yielded by their resin under various circumstances encountered in the plant. To avoid the hazard of facing a short working life,

continuous mixing systems (catalyst and resin) can be very helpful since in this type of arrangement no appreciable amount of time elapses before the mats reach the hot press.

8. *Additive compatibility:* In particleboard manufacture, a number of additives is often incorporated into the board structure to enhance its performance under certain circumstances (see chapter 7). Additives include sizing agents, fire retardants, insect repellents, and fungicides. It is, therefore, essential that the adhesives' properties and curing mechanism not be adversely disturbed when these compounds are added. Furthermore, the efficiency of the adhesive must not suffer with the addition of additives. The current generation of urea- and phenol-formaldehyde adhesives generally meet these requirements.

9. *Flow:* This property refers to the ability of the resin to retain its fluid state in the early moments of hot pressing when the mat has been subjected to intense heat and pressure. A resin with good flow properties continues to wet new particle surfaces and generally permits particle movement in the mat before it is solidified by polymerization. Optimum flow properties allow good board consolidation resulting in acceptable strength and surface characteristics. However, excessive flow is not to be considered desirable because it can result in overpenetration of resin into the particles, thereby depriving the contact area of adequate resin solids.

10. *Bonding efficiency:* Either the board strength obtained per percentage of resin solids used (based on oven-dry wood particle weight) or the amount of resin sprayed per unit particle surface area is an indicator of bonding efficiency. For the sake of economy, the resin must be an efficient binder. The type of resin employed and the techniques of application used are principal factors affecting bonding efficiency. Although the influence of the latter on bonding efficiency is evident, the former can affect efficiency by a number of variables such as the amount used, resin viscosity, solids content, and the temperature during application. For instance, when low solids content is encountered, the resin penetrates too deeply into the particles, thus not fully contributing to bonding in the board structure. Low bonding efficiency can also result when the urea resin generates rapid polymer growth upon contact with the acidic medium (and before full heat and pressure are applied).

11. *Durability:* This property is a measure of the time period during which the adhesive bond will retain its integrity under the given conditions to which particleboard is exposed. The durability of adhesive bonds under

exposure to outdoor conditions is vital in exterior particleboard. The durability of exterior particleboard is measured by the extent of its strength retention, thickness swell, and surface appearance over time. Standard urea resins offer very short durability in exterior situations while the phenolic resins perform a great deal better. Urea resins are durable in most interior applications where excessive levels of moisture and heat are not encountered. Certain types of urea resins have become available which are more durable under high temperature and moisture conditions, compared to standard grades. This resistant generation of ureas offer some advantages over the phenolics, e.g., faster cure rates and easier working properties in addition to their lower cost. Although such urea binders do not offer the long durability required in outdoor situations, they can replace the phenolics in the board core when making a board of layered structure intended for sheltered exterior exposure conditions.[14] A faster rate of cure in the board core allowed by resistant ureas makes long press times unnecessary.

12. *Dilutability:* The resins should be capable of being safely diluted with water. Dilution may be required in some cases to improve the quality of resin blending.

13. *Color:* One of the significant advantages of urea resins is the colorless bond they produce. Not so with phenolics. Generally, it is desirable to use an adhesive which does not impart color to the resulting particleboard.

14. *Cost:* This is, of course, an all-important requirement that all resins must meet. Low cost is essential if any operation is to be economically viable.

Obtaining ideal conditions for many of the properties noted is at present extremely difficult since a large number are at least conflicting if not mutually exclusive. Examples of these properties include reactivity and storage life, early tack loss and low odor, late tack development and early tack loss. It is, therefore, necessary to aim at a compromise in meeting the demands of a given resin user. With further advances in resin technology, it is possible to find ways to optimize many resin properties simultaneously. Optimization has been accomplished for obtaining fast cure rates and good flow properties for both urea and phenolic resins. In resin optimization, an array of factors are relevant:

(a) Wood raw material factors: species, pH, density, moisture content, and bark content. In some cases the resin may be required to tolerate wide differences in species composition of the furnish.

(b) Service requirements: physical and mechanical properties, exposure conditions, cost.
(c) Process particularities: type of particle, level of resin addition, type and amount of additives, forming and processing conditions.

After determining the requirements of a given process, the resin maker has a number of resin manufacturing variables at his disposal to meet those requirements. Some include:

(a) Concentration of reactants in solvent.
(b) Type and number of raw materials.
(c) Ratio of reactants.
(d) Solvent composition.
(e) Type and ratio of hardener.
(f) Type and ratio of buffers.
(g) Time-temperature processing cycle.
(h) Order and rate in addition of ingredients.

Adjustments of various resin characteristics to meet a given set of requirements through modifications in resin formulations and processing is called *tailoring*. There exists a wide latitude in tailoring a resin. Resin properties on which tailoring is often performed are: viscosity, tack, flow, reactivity, pH, buffering, storage stability, formaldehyde release, and additive compatibility. Much tailoring has centered on achieving a fast rate of polymerization in the hot press while maintaining good storage stability and working life. If a given resin lacks the proper balance of working life and cure rate, production efficiency will be hampered. Problems arising include:

(a) Loss of resin activity when the storage of the blended furnish in surge bins is excessive.
(b) Aging or drying of furnish during mat forming and movement to the hot press.
(c) Loss of resin activity from recirculation of over- and underweight mats.
(d) Precure and degradation occurring due to downtime or pressing delays during the movements of mats through the forming line and hot press.
(e) Poor strength properties resulting from brittle, precured surfaces when standard press cycles are used with highly active furnish systems.

These problems have been traced to moisture loss or precure before maximum pressure and heat are applied in the hot press. Tailoring in this case often takes place by modifying the resin with buffers so that it may withstand these effects without excessive impairment of the bonding ability. The art of resin tailoring has been advanced to a stage where:[15]

(a) Tack can appear or disappear at a desired point in the production line to meet a given mat consolidation requirement.
(b) Slow or fast press closing time is accommodated while maintaining fast cure rates.
(c) Resin activity is so adjusted as to tolerate a wide variety of cure conditions and other production particularities.
(d) Viscosity is so adjusted as to generate optimum resin distribution conditions whether the resin is sprayed at room temperature or heated, neat or diluted.
(e) Many storage requirements can be met without undue resin degradation.
(f) Formaldehyde release is kept to a low level.

Experience has shown that there are sufficient differences not only between different board manufacturing processes but also between plants operating on the same process. An adhesive system best suited for a given plant may not be the best choice for the other. Changes in the process such as seasonal effects on the raw material or fluctuations in the dryer efficiency may bring about a need for resin modification.

6–6. Tack

Tack is the adhesive property referring to its ability to adhere to another surface on contact. In other words, tack signifies the stickiness of the adhesive whereby momentary contact with a solid generates an immediate resistance to attempted separation. In a particleboard plant, the term tack is used to describe varying degrees of adhesion between resin-coated particles.

The degree of tack required by different particleboard manufacturing systems for smooth, trouble-free operation, varies. Even within the same process, tack requirements depend on a number of raw material, equipment, and plant environmental factors. For example, particle geometry has a large influence on the degree of tack needed: flakes and fibers, after being formed into mats, generally interlock, thereby holding the mat together until it is consolidated in the hot press. In fact, excessive tack can create problems particularly for fibers.[16] On the other hand, in processes where shavings are used, a high degree of resin tack is desirable since mats made up of such particles tend to crumble around the edges. Also, for mats in which splinters are utilized, high tack is helpful. The wood species, too, has a bearing on the amount of tack required depending on its ability to retain resin droplets on particle surfaces. Species not tending to absorb the resin away from the particle surfaces appear to generate more tack compared to those which allow resin penetration into the body of the particles.

The type of equipment used in a board plant also yields basic information as to the extent of tack needed. In caulless production systems, the mats are depended on to provide a certain degree of rigidity from the time they are formed until the time when they have securely entered the hot press. Mat rigidity generally calls for the resin system to produce a high degree of tack during the given period. High tack can however bring about problematic operation for blending, conveying, and, to some extent, forming equipment. When attrition mills are used simultaneously to reduce and blend fiber furnishes, low tack is highly recommended to avoid difficulties associated with clogging the grinding disks and discharge openings. High tack usually brings about the necessity of frequent clean-ups for conventional blenders equipped with atomizing nozzles, since heavy buildup of resin-coated particles occur. When using high resin tack, the same problem can be encountered with the ductwork in pneumatic discharge systems. A large number of forming systems currently in use in the particleboard industry routinely handle a wide range of resin tack. However, some formers function better when low tack resins are used, producing more uniform mats with insignificant density variations.

Environmental conditions, particularly relative humidity and the corresponding equilibrium moisture content of the furnish, influence resin tack. Should the relative humidity in the plant be such that the material loses moisture after the blending operation, the furnish is likely to face "dry-out," in which residual tack[17] is rapidly lost. Under the reverse circumstance when the particles gain moisture after they are resin-coated, resin tack is likely to survive "dry-out" for longer periods. The influence of temperature on tack is only indirect since such changes affect the relative humidity in the plant.

Operating practices constitute other manufacturing factors with possible direct bearing on resin tack. Diluting the resin will reduce tack and tack retention. Mixing sizing emulsions with the resin could be harmful to resin tack development, particularly in cases when a high level of tack is required. In such cases, the problem may be avoided by separately spraying emulsions perhaps prior to resin blending. It is generally advisable not to modify the resin system once it is received by the board plant unless consulting technical advice indicates otherwise.

Many types of urea-formaldehyde resins supplied to the particleboard industry are so formulated as to develop a moderate degree of tack. This generally suits a large number of users and helps reduce the dust level around such equipment as blenders and mat formers. However, if need be, urea resins can be formulated to develop the required level of tack for a given process. If the plant encounters mat sloughing, sifting, or fracturing and finds that mat surface fines and edges get blown during the final moments of press closing, then raising the tack level may prove helpful in correcting these difficulties. If the mat tends to stick to the prepress rolls, platens or hold-down bars, the factor responsible, however, may be the high level of tack developed by the resin.[18] In summary, to insure trouble-free operation a "tack profile" should be developed for the resin for any given particleboard plant after its processing particularities have been scrutinized.

Phenol-formaldehyde resins, compared to ureas, are less flexible as far as tack is concerned. Further refinements in phenolic resins in this respect is expected, as their use increases in exterior particleboard.

6–7. Formaldehyde Release

In the manufacture of particleboard, formaldehyde release is highly undesirable because it causes various discomforts including itchy, watery eyes. Formaldehyde release, with present know-how, cannot be completely eliminated due to the fact that condensation resins such as urea- and phenol-formaldehyde require an excess of formaldehyde for the reaction process. However, it can be reduced to a minimum through a combination of proper resin formulation, correct manufacturing techniques, and plant environmental controls.

It is now well established that adhesives made with a higher mole ratio of formaldehyde to urea (or phenol) will release a higher level of formaldehyde with other variables remaining constant.[19] Formaldehyde release continues to persist at a high level for a long period of time after the product has begun service in the particular application. Formaldehyde release determined for resin-bonded panels shows that such release drops only by about 35 percent after a storage period of six months. Research on boxes made from particleboard maintained at 25°C. and 45 percent relative humidity indicates that formaldehyde concentration in the interior of such boxes is well in excess of 6 milligrams per cubic meter of air.[20] Such concentrations severely limit the use of these boxes. A check of boxes stored for some seven weeks shows that formaldehyde concentration drops only by 10 to 20 percent on the basis of freshly manufactured levels. Ventilating and airing the boxes is of little use: after three days of ventilation, once the boxes are sealed again, formaldehyde concentration is quickly restored to preventilation levels.

6-5. *Effect of ammonium chloride on formaldehyde release.*

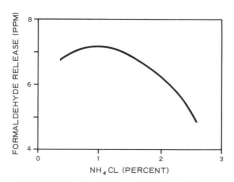

SOURCE: Deppe, H. J., and Ernst, H. 1966. On reducing compacting time in chipboard manufacture. Redrawn in modified form by S.I.U. Cartographic Laboratory

A high molar ratio of formaldehyde to urea has the disadvantage of a greater degree of formaldehyde release, but it imparts certain desirable properties, e.g., increased stability in solution and infinite true solubility, among other advantages already pointed out. A compromise is thus mandatory. To reduce formaldehyde release, the addition of a limited amount of urea to the resin solution just prior to spraying the particles may be helpful.

The type and amount of hardeners used in the resin formulation also have a substantial influence on formaldehyde release. Figure 6–5 illustrates that the addition of 2 percent or more of ammonium chloride (NH_4Cl) reduces formaldehyde release. The reduction with increasing ammonium chloride is due to combining of $-NH_4$ groups with formaldehyde to produce hexamethylenetetramine. Increasing the proportion of buffering ammonia also reduces formaldehyde release due to the same reason. However, hardeners and buffers, as noted earlier, cannot be increased beyond a compromise limit without disadvantageous consequences on other resin properties. Excessive levels of hardener will increase the danger of short storage life and increased precure. Increased buffers, beyond an optimum limit, endanger the resin's fast cure rate thereby adversely influencing the strength and dimensional properties of the board.

Formaldehyde release is further influenced by the moisture content of the furnish and the length of the hot pressing operation. Figure 6–6 illustrates that the amount of water evaporated per unit surface area of the mat during the hot pressing has a bearing on the amount of formaldehyde released. This figure was obtained by varying the moisture content of the furnish while maintaining other relevant variables constant. Based on information provided by figure 6–6, it is noted that formaldehyde release is reduced as the amount of water evaporated per unit surface area of the mat increases. Thus, plants operating with low levels of furnish moisture, other factors being the same, can expect a greater formaldehyde odor problem. Again, one cannot reduce furnish moisture below a compromise limit without adversely sacrificing desirable board properties. Plath,[21] in connection with furnish moisture, contends that augmentation of surface layer moisture results in an increase in the level of formaldehyde release, whereas the alteration of the moisture content in the middle layers is of small influence.

The period of time during which the board remains in the hot press has an appreciable influence on formaldehyde release. Tests show that a longer press time results in reductions in the amounts of formaldehyde released by the boards (fig. 6–6). It is noted from this figure that for the particular resin used, a press time of approximately four minutes or longer results in a formaldehyde release of less than ten ppm (parts per million). Lengthening the time, however, has its obvious drawbacks on production. Reductions in relative humidity and temperature of inplant environment also appear to decrease formaldehyde release by a given resin.

In summary, it is realized that no single action affecting formaldehyde release is normally taken to reduce the odor of formaldehyde. A combination of factors such as the type of resin, type and amount of hardeners

6–6. *Effect of water evaporated from the mat and length of press time on formaldehyde release.*

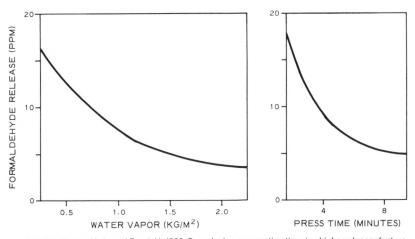

SOURCE: Deppe, H. J., and Ernst, H. 1966. On reducing compacting time in chipboard manufacture. Redrawn in modified form by S.I.U. Cartographic Laboratory

and buffers, as well as manufacturing variables are pooled together to bring the odor release to an acceptably low level. Finishing particleboard with formaldehyde-binding coatings has been recommended[22] in circumstances where economics so permit and where the formaldehyde odor is highly objectionable.

To determine the extent of formaldehyde release, various techniques have been employed. Formaldehyde given off by consolidated panels has been absorbed in hydroxylamine hydrochloride solution followed by titrating the hydrochloric acid liberated. Another technique involves the diffusion of formaldehyde from the boards into a collecting agent. Treatment of the collected formaldehyde with chromotropic acid develops a color, the depth of which is measured by a spectrophotometer in the visible range. The absorbance thus specified is related to the quantity of formaldehyde present by means of a previously determined standard curve.

6-8. Influence on Board Properties

Resin costs constitute a major production expense. Returns in terms of board properties from every unit of the resin must be maximized. The amount of resin used in board manufacture is expressed in one of two ways:

(a) Percent of resin solids based on the oven-dry weight of wood furnish.
(b) Amount of resin solids used (in grams or pounds) per unit surface area of the particles.[23]

The latter method of expressing the amount of resin in boards is more meaningful because it takes into consideration the surface area of the particles. However, expressing the resin usage in percent, based on oven-dry furnish weight, is prevalent in industry, due to the ease in applying and comprehending the technique. In any event, one method of resin measurement can be converted into the other by using the following relationship:

$$A' S = C \qquad\qquad (6-1)$$

in which A' represents the particle surface area in m^2 per gm of oven-dry furnish (or 1000 ft.^{2} per pound), S is the dry resin spread in gm. per m^2 of wood particle surface area (or pound per 1000 ft.^{2}) and C is the binder consumption in gm. per 100 gm. of oven-dry wood furnish (or pound per pound).

The amount of resin used in the manufacture of particleboard varies over a considerably wide range. It may vary from zero to 15 percent. Most often, however, the amount of resin used under current operating conditions ranges from 5 to 9 percent for urea resins and 3 to 6 percent for phenolics. Generally, boards intended for such uses as floor underlayment contain less resin compared to those used in furniture industry. For furniture use, urea-bonded resin usage may average 7 to 9 percent for single-layer boards. When multi-layer particleboard is intended for furniture manufacture, the resin usage in face layers may be as high as 10–12 percent, with 6 percent or higher in the core. Using fiber furnishes, urea resin content may be as high as 8 percent or more in order to obtain adequately high internal bond strength. Using resin quantities of less than 3 percent can create a difficult task in attempting to obtain proper distribution: the amount of resin becomes too small to achieve good resin coverage of the furnish. This problem is particularly acute when low-density, thin particles, i.e., flakes, are involved due to the large surface area per unit weight.

In certain unusual circumstances, the use of external adhesives has not been found necessary. Vakhrusheva and Petri,[24] took advantage of natural binding material present in some species such as European larch to produce a well-consolidated particleboard possessing adequate mechanical and dimensional properties. The authors working with larch (containing some 12 percent soluble polysaccharides—arabogalactan gum) believe that a portion of the gum present in the wood species itself can be hydrolyzed under pressure and heat to produce sugars. The sugars then interact with the activated lignin formed during pressing as a result of the splitting up of the carbon-lignin complex. This is followed by polycondensation, which splits off water leading to the formation of compounds capable of bonding the particles.

In the manufacture of multi-layer boards, the resin amount can vary in different board layers so that a desired set of board properties is obtained. In most cases, a higher percentage of resin is added to the face layers in order to obtain good flexural and surface appearance properties. In such cases, the magnitude of the resultant resin content may be of interest from the standpoint of resin usage. If we let R_r represent the resultant resin content, R_c and R_f the resin contents in the core and face layers of a three-layer board respectively, we can then write the following relationship:[25]

$$R_r = \frac{R_c g_c (t - 2s) + R_f g_f \cdot 2s}{g_c (t - 2s) + g_f \cdot 2s} \tag{6-2}$$

in which g_c and g_f refer to the dry density of core and face layers respectively, t is the overall thickness of the board, and s is the face layer thickness. It is noted that equation 6–2 can easily be expanded for other board constructions in which more than three layers are involved. Continuing our analysis of a three-layer board, we can rewrite equation 6–2 as follows:

$$R_r g_r = R_c g_c (1-\lambda) + R_f g_f \cdot \lambda \tag{6–3}$$

in which $\lambda = \dfrac{2s}{t}$ is the shelling ratio and g_r is the resultant density of the board expressed as follows:

$$g_r = \frac{g_c(t-2s) + g_f \cdot 2s}{t}.$$

Equation 6–3 can take the following form:

$$R_r = \frac{R_c}{1+m} + \frac{mR_c}{1+m} \tag{6–4}$$

where

$$m = \frac{\lambda g_f}{(1-\lambda)g_c}$$

is a density ratio in which g_f represents the surface layer contribution and $(1-\lambda)g_c$ is the weight of the core layer. Taking into consideration the compression ratio $C = \dfrac{g_f}{g_c}$ and the resin ratio $R = \dfrac{R_f}{R_c}$ then the resultant resin content can be calculated by:

$$R_r = R_c \frac{1 + \lambda(CR-1)}{1 + \lambda(C-1)} \tag{6–5}$$

which indicates that in order to obtain low overall resin usage, low core resin content but higher R and C should be considered.

Bending strength appears to be enhanced in a linear fashion as the percent of resin is increased.[26] Figures 6–7[27] and 6–8[28] show two examples for beech flakes and Douglas-fir planer shavings respectively. Figure 6–7 shows a set of four lines each for a different level of density based on statistical equations fitted to empirical data. This figure indicates that a 2 percent urea resin increase over the range shown brings about a 25 kg/cm² increase in bending strength.

Increasing resin content also appears to increase internal bond strength in a linear fashion, as figures 6–7 and 6–8 for beech flakes and Douglas-fir shavings respectively indicate. Thus, in a manufacturing operation in

6–7. *Effect of resin content on strength and thickness swelling properties when a range of particleboard densities are considered. All boards were made of European beech flakes.*

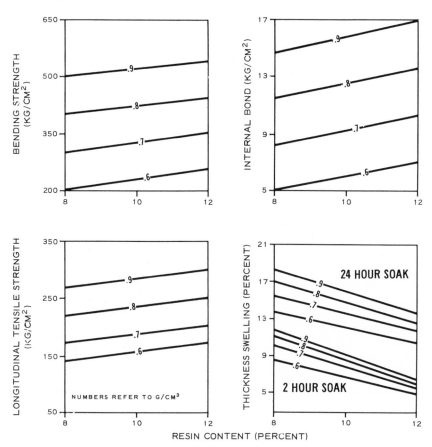

SOURCE: Stegmann, G., and Durst, J. 1964. Beech pressboard.
Redrawn in modified form by S.I.U. Cartographic Laboratory

which the board resin content (or that of the core layer in multi-layered constructions) is reduced, keeping all other variables constant, a drop in this property is to be expected. The tensile strength parallel to the board surface is also influenced by varying the resin content, as the example shown in figure 6–7 illustrates. In addition, screw withdrawal resistance (an important property in many particleboard applications) appears to be influenced in a generally positive manner by the increasing resin content.

6–8. *Effect of resin content on strength of Douglas-fir particleboard made with shavings at three different densities.*

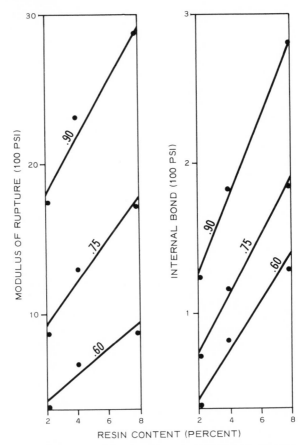

SOURCE: Talbott, J. W., and Maloney, T. M. 1957. Effect of several production variables on the modulus of rupture and internal bond strength of boards made of green Douglas-fir planer shavings. Redrawn in modified form by S.I.U. Cartographic Laboratory

The dimensional stability, particularly the extent of thickness swelling exhibited by the board, has been of intense interest since particleboard was introduced to the market. Thickness stability during changes in environmental humidity is a basic requirement. The extent of swelling is usually measured after the boards have been soaked in water at room temperature (70°F.) for periods of two and twenty-four hours. Figure 6–7 shows that increased resin content results in more thickness stability.

Linear dimensional changes behave in a similar fashion although the magnitude of linear change is considerably less. Obviously, the extent of thickness expansion (or linear expansion) under two-hour soak will not be as large as the twenty-four-hour soak since, in the former, water penetration into the board structure will be very incomplete.

The development of the exterior particleboard has brought under attention the question of binder durability under outdoor exposure conditions. An appreciable amount of work has been performed to find ways of manufacturing a truly exterior-grade particleboard capable of under-

6–9. *The effect of test fence exposure time on particleboard properties with resin content percentages and resin types indicated.*

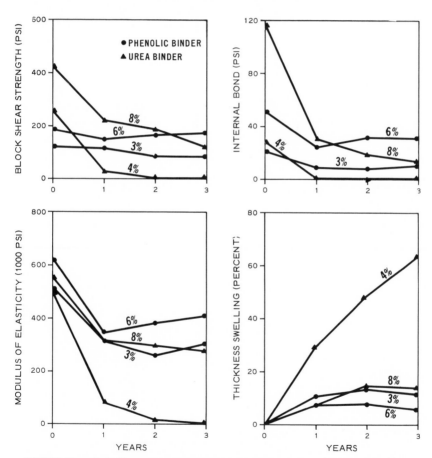

SOURCE: Hann, R. A., Black, J. M., and Blomquist, R. F. 1962. How durable is particleboard. Redrawn in modified form by S.I.U. Cartographic Laboratory

going rigors of the service environment for which it is intended. Tests at the United States Forest Products Laboratory in Madison, Wisconsin, have shown that the level of resin in the under-9-percent range has an appreciable effect on strength retention and thickness swell of medium-density particleboard exposed to the outdoors over a three-year period.[29] For instance, figure 6–9 shows that the extent of strength retention is substantially improved as the resin level is elevated from 4 to 8 percent for urea-formaldehyde and from 3 to 6 percent for phenol-formaldehyde. The board containing only 4 percent urea-formaldehyde loses practically all of its strength after a one-year exposure period. The particleboard containing 8 percent urea resin retains some of its strength, particularly its modulus of elasticity, after it is exposed outdoors for a three-year period.

Thickness swelling is reduced as the quantity of resin in the board increases (fig. 6–9). The phenolic-bonded board shows better performance per percent resin solids added compared to urea-formaldehyde bonded particleboard. The dramatic difference between the two types of resins is usually brought out when boards are subjected to the ASTM Accelerated Aging Exposure tests as table 6–1 illustrates. Urea-bonded boards disintegrate after only one cycle of exposure to this test regardless of whether the boards contain 4 or 8 percent resin. The phenolic-bonded boards generally retained a considerable percentage of their original strength upon completion of all cycles in the test.[30] "Exterior grade" boards currently made by the industry contain 3–10 percent phenol-formaldehyde binder.

6–1. *Strength properties of urea and phenolic-bonded particleboards after ASTM Accelerated Aging exposure.**

Binder Type	Percent Resin Added	Density pcf	Modulus of Rupture psi	Modulus of Elasticity 10^3 psi	Internal Bond psi	Percent of Retained Strength		
						Modulus of Rupture	Modulus of Elasticity	Internal Bond
Urea	4	34	0	0	0	0	0	0
Urea	8	35	0	0	0	0	0	0
Phenolic	3	34	1230	231	1	46	45	5
Phenolic	6	35	2510	381	31	68	62	61

* Bending data are average of three specimens; internal bond average of two.

SOURCE: Hann, R. A., Black, J. M., and Blomquist, R. F. 1962. How durable is particleboard.

Other board properties of importance affected by the level of resin present in the board are surface appearance and machinability. A smoother surface appearance will result as the resin content of the board (or the surface layers) is increased. The establishment of sound relationships between surface appearance and other raw material and manufacturing variables is difficult at present due to the absence of a simple and quantitative surface test procedure. All machining characteristics of particleboard appear to improve as the resin content is increased.

6–9. Determination of Resin Content in Boards

In the manufacture of particleboard, the need may arise to determine the actual resin content in the products made. Such determinations may be associated with quality control work, checking the integrity of operation of resin metering and blending equipment, and the uniformity of operation over time. The need for resin content determination in products made may also arise in connection with research work.

Urea-formaldehyde: In determining the urea resin content of the boards, advantage is taken of the fact that this amino resin has a distinct element in its chemical makeup—nitrogen. The nitrogen content of the resin is determined and subsequently related to the total resin solids content of particleboard.[31] Standard techniques of determining the nitrogen content, e.g., Kjeldahl or Dumas methods, can be used to specify indirectly the amount of urea resin present. However, improved techniques are now available providing a more rapid[32] and reproducible means compared to the standard techniques of nitrogen analysis just noted. Stegmann and Ginzel[33] have proposed a refined technique of urea resin content determination originally suggested by Klauditz and Meier.[34] It involves the use of perhydrol and sulfuric acid (to disintegrate the wood portion of the ground particleboard sample) and the Kjeldahl method of nitrogen analysis.

In the technique proposed by Stegmann and Ginzel, two grams of finely ground urea resin-bonded particleboard is poured into a 300 ml. wide-neck Erlenmeyer flask which already contains two parts of perhydrol and one part of concentrated sulfuric acid (by volume). Upon addition of the ground material, the mixture is heated under a hood on a mild flame, whereupon an intense reaction occurs. Heating continues with progressively stronger flames until sulfuric acid vaporizes and escapes. Should dark coloring appear at this stage, it would indicate that the oxidation of wood is not fully completed. In this case, the flask is allowed to cool followed by the addition of one ml. of perhydrol and the reheating of the flask. If the solution has still not turned clear, the procedure of adding perhydrol is repeated a number of times, if necessary, until the solution color disappears. Should any iron be present in the solution, a slight permanent yellowing can occur due to the presence of ferric sulfate. After the clear solution is obtained, it is cooled, followed by addition of about 0.5 to 1.0 gram of solid oxalic acid. This is then heated on a low flame until the oxalic acid is dissolved. The heating continues until the foaming disappears and the resulting sulfuric acid vapors escape, upon which the mixture is once again cooled. At this time, the mixture is diluted with approximately 100 ml. of distilled water—with care exercised to rinse the walls of the flask so that any sulfuric acid which may be clinging on the flask walls disappears. The diluted solution is neutralized by adding 10 percent caustic soda lye in the presence of methyl red while the flask is thoroughly cooled in an ice-filled container. The cooling prevents possible escape of small amounts of ammonia. This escape is permitted when localized heating occurs thereby making the nitrogen determination slightly inaccurate. Once the solution is neutralized (the neutral point is the n set at n/10 NaOH), 10 ml. of 40 percent formaldehyde is added to react with the ammonium sulfate present in the solution resulting in sulfuric acid liberation:

$$2(NH_4)_2SO_4 + 6HCHO \rightarrow (CH_2)_6N_4 + 2H_2SO_4 + 6H_2O$$

To titrate the newly generated sulfuric acid, n/5 NaOH is employed in the presence of phenolphthalene serving as an indicator. The last ml. of the caustic soda at this stage should be added in drops so that a sudden change is easily observed. The reaction proceeds at a slower pace towards the end in the titration process.

Since formic acid is always present in formaldehyde, a 10 ml. of the latter must, therefore, be titrated with n/5 NaOH in the presence of phenolphthalene. The amount of lye used in ml. must be subtracted from that employed in the titration process when determining the nitrogen content. The latter can now be determined as follows:

$$q = .0028 \times 10^4 \frac{s}{pp'} \tag{6-6}$$

in which q represents the percent of N_2 in the dry material, s is the amount of n/5 NaOH in ml., p is the quantity of ground material in grams, and p' is its dry content in percent. The constant, 0.0028, used in the above equation is due to the fact that every ml. of n/5 NaOH used signifies the presence of 0.0028 gram of N_2.

To relate the nitrogen content to the amount of urea resin present in the board, a number of parameters related to the resin and the hardener must be known. These are:

(a) Nitrogen content of the liquid urea resin, q_1, in percent.
(b) Nitrogen content of the liquid hardener, h_1, in percent.
(c) Solids content of the liquid resin, m_1, in percent.
(d) Solids content of the liquid hardener, n_1, in percent.
(e) Hardener to adhesive ratio, a, used in manufacturing the board.

Among the parameters mentioned, m_1 and n_1 are normally supplied by the resin maker (or can easily be determined in the laboratory), and a is known in the course of particleboard manufacture. The nitrogen contents of the urea resin and the hardener, namely q_1 and h_1, can be determined as follows:

For the resin, a 0.25 gram sample of the liquid is placed in the 300 ml. Erlenmeyer flask to which 20 ml. of the perhydrol-sulfuric acid mixture (two parts perhydrol to one part concentrated sulfuric acid by volume) is added. Then the procedure for nitrogen determination is followed as described above. Also, for the hardener, a 0.25 gram sample is taken from the liquid and placed in the 300 ml. Erlenmeyer flask. This time the hardener sample is evaporated in a water bath until all the ammonia is removed, followed by cooling. Again, a 20 ml. of the perhydrol-sulfuric acid is added and the nitrogen determination performed in the same manner.

Once the resin and hardener parameters are known, the conversion factor of the nitrogen to actual resin content can be arithmetically accomplished. The nitrogen content of the resin, after the hardener has been added, n, is:

$$n = q_1 + ah_1 \qquad (6-7)$$

Basing n on solids content of the resin, we get:

$$n' = 100\frac{n}{m_1} \qquad (6-8)$$

in which n' represents the amount of N_2 present in 100 parts of solid resin. Thus, the ratio of solid resin to nitrogen content f will then be:

$$f = \frac{100}{n'} \qquad (6-9)$$

by which the determined nitrogen content must be multiplied to arrive at the urea resin content used in the board. As an example, let us assume that the solids content of urea resin being used is 65 percent and that of the hardener is 44 percent. Furthermore, let us assume that the liquid resin has a nitrogen content of 18 percent and the liquid hardener possesses a nitrogen content of 11 percent. If we are adding 10 parts of hardener to 100 parts of urea resin in making a given board, the nitrogen content of the resin-hardener liquid will be boosted as shown in equation 6–7. In our example this will be:

$$18 + 1.1 = 19.1 \; N_2.$$

To obtain the nitrogen content based on the solids content of the resin, equation 6–8 will give $19.1/65 \times 100 = 29.38 \; N_2$ in 100 parts of solid resin. Thus, f in our example will be $f = 100/29.38 = 3.40$.

In calculation of the nitrogen content in particleboard, it is advisable to consider the nitrogen content present in the wood species. The nitrogen content of the wood species should be subtracted from the nitrogen determination results before it is multiplied by f to arrive at the resin content present in the board. This can normally be accomplished by testing the wood used in a given operation separately and periodically, depending upon whether or not significant changes have occurred in the wood raw material makeup. The nitrogen contents of wood species used in particleboard manufacture may range from zero to 2 percent or higher. Thus, if for instance, we determine that there is a nitrogen content of 2.10 percent in the board sample with the wood species (or species mix) containing 0.15 percent nitrogen content, the actual nitrogen attributable to the urea resin will then be only 1.95 percent. Assuming we have calculated a conversion factor, f of 3.40, then the actual resin content will amount to $1.95 \times 3.40 = 6.63$ percent.

Phenol-formaldehyde: The determination of this resin is more difficult compared to urea-formaldehyde. It is strongly resistant to chemical attack and contains no distinct element such as nitrogen in its chemical composition whose quantity would be proportional to the amount of resin solids. Thus, in attempts to isolate phenol-formaldehyde in particleboard, highly reactive chemicals such as molten β-napthol[35] and chlorosulfonic acid[36] have been utilized. Extreme caution in handling these chemicals as well as long operation time constitute significant disadvantages of phenolic resin determinations.

Ettling and Adams[37] utilized chlorosulfonic acid to determine the phenolic resin content in particleboard. In their procedure, a small quantity of shaven or ground phenolic-bonded particleboard is first oven-dried at 100°C. to a stable weight. Then, one-to-five-gram individual samples of the oven-dry material is placed in a suitable container to which chlorosulfonic acid is added at the rate of five grams of acid per one gram of ground sample. The mixture is subsequently left alone at room temperature for approximately one hour so that the wood component of the sample is dissolved by the acid present. The mixture is then filtered through tared fritted glass funnels, washing the residue with isopropanol. The procedure continues with drying the funnels at 150°C. for an hour, after which, they are cooled and weighed to yield the amount of phenol-formaldehyde resin solids used in the ground sample. Ettling and Adams contend that this technique is accurate to within $+2$ percent of the actual amount of resin present. However, due to disadvantages pointed out above and the fact that the method fails for phenolic resins cured with ammonium hydroxide,[38] it cannot be considered an ideal solution for determining phenolic resins.

Another technique radically different from that just described, is the one proposed by Chow and Mukai[39] in which the authors utilize a combination of standard transmission and differential techniques of infrared spectrometry to determine the phenolic resin content. In this method, $2 + 0.2$ milligrams ground samples (passing through 140-mesh screen) are used to measure the absorbance of the skeletal vibrations of the phenolic ring at 1470–1500 cm^{-1} band taken as the quantitative absorption by the resin and 2900 cm^{-1} taken as the absorption of both wood and the resin present in the sample. The phenolic resin content in the sample would then correlate with the absorbance ratio between these two bands just indicated. Figure 6–10 shows the phenol-formaldehyde

6–10. *Calibration curves for phenol-formaldehyde using aspen.*

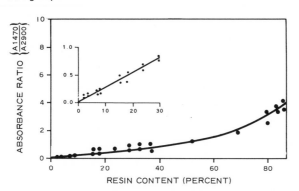

SOURCE: Chow, S. Z., and Mukai, H. N. 1969. An infrared method of determining phenol-formaldehyde resin content in fiber and wood product.
Redrawn in modified form by S.I.U. Cartographic Laboratory

resin content—absorbance ratio relationship determined for an aspen particleboard. Chow and Mukai state that the accuracy of this technique is within 5 to 10 percent of the resin's solid content present in the sample. However, there are drawbacks:

(a) Extreme care in measurements must be exercised to filter out noise in the instrumentation, particularly when low resin levels are used in the sample under examination.

(b) A separate calibration curve similar to that presented in figure 6–10 must be constructed for each resin and degree of cure induced (due to the possible dependence of the absorptivity on the molecular weight of the resin). Thus, this technique, too, cannot adequately fill the need for an accurate, simple way of determining the phenol-formaldehyde resin content in boards.

Further research in finding a simple, reasonably accurate technique is clearly needed.

Notes

1. Urea-formaldehyde adhesives are thermosetting which implies that polymerization (curing) can occur upon heating.

2. Megson, J. J. L. 1958. Phenolic resin chemistry.

3. Carroll, M. N., and McVey, D. 1962. An analysis of resin efficiency in particleboard.

4. Laidlow, R. A., and Hudson, R. W. 1969. Tour of chipboard factories and research institutes in Europe with special reference to the production of exterior quality boards.

5. Hann, R. A., Black, J. M., and Blomquist, R. F. 1962. How durable is particleboard.

6. Liiri, O. 1969. The most recent developments in the wood particleboard line.

7. Laidlow, R. A., and Hudson, R. W. 1969. Tour of chipboard factories and research institutes in Europe with special reference to the production of exterior quality boards.

8. Hall, R. B., Leonard, J. H., and Nicholls, G. A. 1960. Bonding particleboard with bark extracts.

9. Anderson, A. B., Breuer, R. J., and Nicholls, G. A. 1961. Bonding particleboards with bark extracts.

10. Plomley, K. F. 1966. Tannin formaldehyde adhesives for wood. Part 2: Wattle tannin adhesives.

11. Based on data supplied by the resin maker.

12. Anon. 1967. Bulk handling of liquid resin glue.

13. Lambuth, A. L. 1967. Performance characteristics of particleboard resin binders.

14. Ibid.

15. Ibid.

16. Raddin, H. A. 1967. High frequency pressing and medium density board.

17. Residual tack refers to the degree of tack retained by a dried film of resin.

18. Lambuth, A. L. 1967. Performance characteristics of particleboard resin binders.

19. Berger, V. 1964. Urea binders with low content of free formaldehyde.

Miels, G., and Scheibert, W. 1957. Shortening of compacting time.

Mohr, E. 1964. The metal changing plate in particleboard plant.

Starzyska, K. 1964. Urea binders with low content of free formaldehyde, their properties and application.

20. Deppe, H. J., and Ernst, H. 1966. On reducing compacting time in chipboard manufacture.

21. Plath, L. 1967. Tests on formaldehyde liberation from particleboard. Part 4: Effect of moisture content in chip mat on formaldehyde liberation.

22. Wittmann, O. 1962. The subsequent dissociation of formaldehyde from particleboard.

23. The amount of resin added to particleboard usually lies in the 4–12 gm/m² range based on dry resin weight.

24. Vakhrusheva, I. A., and Petri, V. N. 1964. Use of comminuted larch wood for making plastics without binders.

25. Keylwerth, R. 1958. On the mechanics of the multi-layer particleboard.

26. Burrows, C. H. 1961. Some factors affecting resin efficiency in flakeboard.

Stegmann, G., and Durst, J. 1964. Beech pressboard.

Talbott, J. W., and Maloney, T. M. 1957. Effect of several production variables on the modulus of rupture and internal bond strength of boards made of green Douglas-fir planer shavings.

27. Flake dimensions were: $t = 0.17$ mm., $w = 2$ mm., and $l = 15$–20 mm. Other manufacturing data: Total press time $3\frac{1}{2}$–4 minutes; board thickness, 8 mm.; press temperature, 170°C. at pressure of 19.5–33 kg/cm² depending on board density desired. Urea resin used.

28. Planer shavings utilized were passed through 4-mesh screen but held on 16-mesh $(-4+16)$.

29. Hann, R. A., Black, J. M., and Blomquist, R. F. 1962. How durable is particleboard.

30. West Coast Adhesive Manufacturers Association (WCAMA) has put forth a more simplified test than that described by ASTM-1037 which it provides similar predictive values of the outdoor exposure performance of particleboard to those obtained from the ASTM test. For details the reader is referred to:

West Coast Adhesive Manufacturers Association. 1970. Accelerated aging of phenolic resin-bonded particleboard.

West Coast Adhesive Manufacturers Association. 1966. A proposed new test for accelerated aging of phenolic resin-bonded particleboard.

31. Klauditz, W. 1954. Determination of urea-formaldehyde resin content (adhesive) in wood particleboards.

Pinion, L. C. 1967. Evaluation of urea-formaldehyde and melamine resins in particleboard.

Pinion, L. C., and Farmer, R. H. 1958. Wet combustion of organic materials.

Seifert, K. 1959. The analysis of wood particleboards.

32. About 20 minutes as compared to 5 hours for the straight Kjeldahl analysis.

33. Stegmann, G., and Ginzel, W. 1965. Determination of the content of urea-formaldehyde adhesives in particleboard.

34. Klauditz, W., and Meier, K. 1960. Determination of the percentage of urea and melamine resins in wood particleboards.

35. Sciconolfi, C. A. 1960. Resin solids in a saturated sheet.

36. Ettling, B. F., and Adams, M. F. 1966. Quantitative determination of phenolic resins in particleboard.

37. Ibid.

38. Chlorosulfonic acid forms salts with the basic nitrogen atoms which are incorporated into phenolic resins when such resins are cured with ammonium hydroxide or hexamethylenetetramine.

39. Chow, S. Z., and Mukai, H. N. 1969. An infrared method of determining phenol-formaldehyde resin content in fiber and wood product.

7

Additives

7-1. General

Additives refer to chemical compounds incorporated into particleboard in the course of manufacture. Current additives are added to obtain better properties in one or more of such areas as water repellency, fire retardancy, and resistance to fungal and insect attack. Hardeners and buffers, often added to the resin system, also fall under the definition of additives.

7-2. Wax Sizing

Wax sizing, currently the most prevalent additive, is designed to confer a degree of *liquid water* repellency upon the board. The practice of incorporating wax into particleboard and molding is done whether urea on phenolic binders are used. The function of wax sizing is not limited to water repellency alone: wax provides the furnish with slip characteristics, thereby helping to reduce obstruction and maintenance in forming stations, blow pipes, blenders, cauls, and conveyor belts. Furthermore, the use of wax avoids overpenetration in subsequent gluing (or surface treatment materials) of particleboard for overlays.[1] Drawbacks, however, have been pointed out[2] and blamed on the presence of small, oily substances in wax. Oily substances are believed to create difficulties in secondary gluing, priming, and painting of particleboard surfaces.

7–1. *Thickness swelling in particleboard made of three species and water-soaked for an extended period of days.*

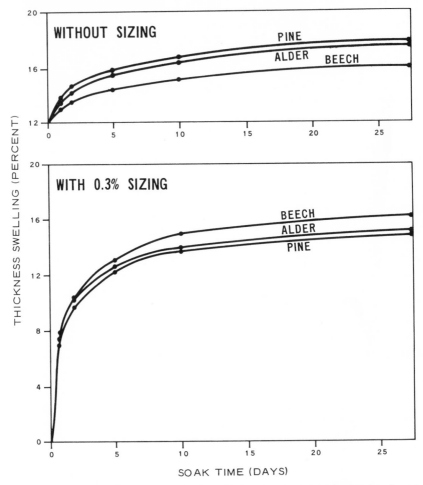

SOURCE: Kehr, E. 1964. Investigation of the suitability of various types and categories of wood for the manufacture of particleboard.
Redrawn in modified form by S.I.U. Cartographic Laboratory

Wax sizing does not offer any protection against *water vapor* changes in the surrounding environment. Wax addition, therefore, can help exterior particleboard repel the water to which it may be subjected. The water repellency afforded by wax addition is generally short-term.[3] Figure 7–1 shows that when particleboard is subjected to liquid water at room temperature, for a number of days, the differences between wax-treated samples and non-treated controls essentially disappear in a relatively short time. Thus, wax sizing provides protection against short-term accidental wetting. To obtain protection for board surfaces, a number of European plants incorporate wax sizing only into the surface layers of urea-bonded panels.[4] These plants favor the addition of wax to all phenolic-bonded boards because such products may become subjected to repeated short-time wetting.

Water repellency provided by wax sizing is primarily due to reduction of capillary flow in particleboard. In reality there are two types of capillaries in particleboard: cell lumen openings and the voids between particles. Wax, due to its large molecular size, cannot enter the former. Its action is strictly limited to the latter type[5] composed of relatively small, irregularly shaped capillaries. Wax sizing repels liquid water by making the particle surfaces surrounding such capillaries less wettable. When water is capable of easily wetting the surfaces, it can rise easily in the capillary and thereby readily penetrate the board. Another important factor concerns the rate with which liquid water is capable of penetrating the board structure since this determines how fast the products respond to water contact. This rate is a function of the capillary radius, pressure, and depth of penetration. Although a decrease in the radius of the capillary in the board structure will tend to facilitate a deeper water penetration, the rate of water flow decreases due to the pressure of even greater frictional forces. When, for instance, the overall particleboard density is increased, resulting in smaller inter-particle voids, a progressively lesser amount of water is admitted into the board structure over a given period of time. In reality, the true case of water penetration into particleboard involves a more complex set of relations including intercapillary dependencies. For instance, observing penetration under a microscope, Albrecht[6] believes that particle-particle bonds interfere with the simple capillary flow in the voids as the water continues to penetrate deeper into the board structure.

The amount of wax added to the board varies from plant to plant but is generally limited to 1 percent or less, based on wax solids and the oven-dry weight of the wood furnish. The point of "diminishing returns" is quickly reached beyond this limit as indicated in figure 7–2. These figures show that gains in thickness swelling and water absorption—after immersing the particleboard in water for a twenty-four-hour period—begin to level off once the wax addition reaches approximately 0.6 to 0.8 percent.[7] For different manufacturing and raw material conditions, the optimum limit of wax addition may vary somewhat from those presented. But, it is believed that wax additions over 1 percent will not bring about further worthwhile improvements in dimensional stability and water

7–2. *The effect of the percentage of wax sizing on water absorption and thickness swelling of three particleboards each made with a different species.*

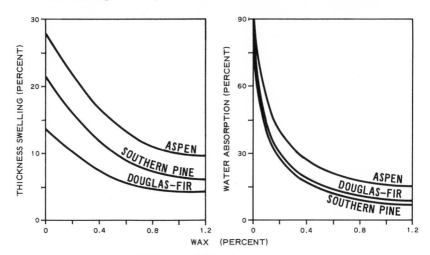

SOURCE: Heebink, B. G. 1967. Wax in particleboards.
Redrawn in modified form by S.I.U. Cartographic Laboratory

absorption properties. Excessive wax additions are likely to cause reductions in board strength.

Stegmann and Durst,[8] working with beech particleboard, found that wax sizing not only reduces dimensional instability in regard to liquid water contact, but it also causes drops in bending, internal bond, and longitudinal tensile strengths. Furthermore, the use of wax emulsions from three different sources resulted in differences, sometimes considerable, in board properties obtained. Figure 7–3 shows, for instance, that strength reductions even below the 1 percent range can result by the addition of wax sizing. Sizable differences in wax emulsions obtained from various sources are indicated: wax from source C significantly interfered with bonding, thereby sharply reducing the strength of the resultant board. On the other hand, wax from source C exhibited a better performance in providing the boards with resistance to water absorption and the accompanying thickness swelling.[9]

Paraffin wax, which is normally used in particleboard is usually a colorless or white substance consisting of a mixture of solid hydrocarbons. It is a by-product of the petroleum distillation process. In such hydrocarbons, the carbon atoms are linked together by single bonds with hydrogen atoms satisfying the remaining valences. Wax, a translucent mass exhibiting a crystalline structure, has no odor or taste and appears slightly greasy to the touch. Its melting point ranges from 115° to 190° F. There

7–3. *The effect of the source from which wax sizing is obtained on a number of beech particleboard properties. The letters A, B, and C refer to the commercial sources.*

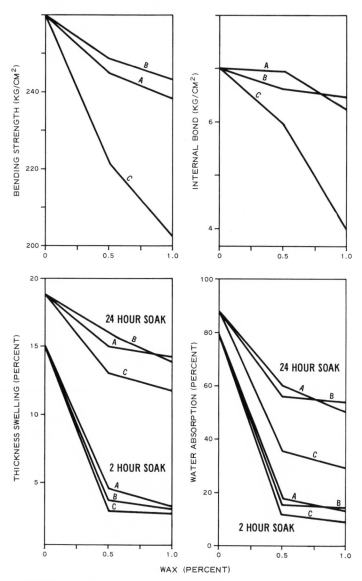

SOURCE: Stegmann, G., and Durst, J. 1964. Beech pressboard.
Redrawn in modified form by S.I.U. Cartographic Laboratory

are different grades of wax with a variety of properties available in the industry. Small amounts of liquid impurities in the form of oil are often present in industrial waxes. The wax distillate fraction of crude petroleum produces paraffin wax which is subsequently separated from other low-melting waxes. Figure 7–4 schematically outlines how various wax products are derived. Distillation of crude petroleum produces gases, gasoline, fuel oil, paraffin distillate, and residue. The paraffin distillate so obtained provides the major source of wax for use in the manufacture of particle-board. Also, the microcrystalline waxes derived from the residue are sometimes employed for this purpose. Wax sizing is incorporated into the particle furnish in one of two forms: wax emulsions and molten waxes. The former currently enjoys preference over the molten varieties because a better wax distribution over the particle surface area can be obtained. This is an important factor since, as pointed out earlier, very small quantities of wax are normally involved in the sizing operation.

A variety of emulsification systems is used for waxes which are basically categorized as anionic, cationic, and nonionic (neutral). Such systems colloidally suspend wax particles of approximately one micron in diameter in water and envelope them with the emulsifying agent.[10] The type of emulsifier utilized is dependent upon such factors as conditions of resin compatibility, mechanical stability, and cost. The majority of

7–4. *Crude oil distillation as related to the production of wax.*

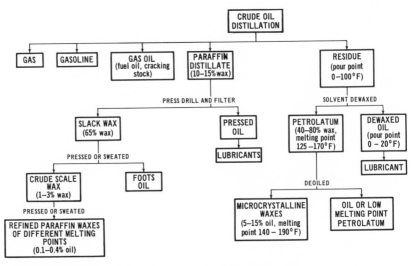

SOURCE: Albrecht, J. W. 1968. The use of wax emulsions in particleboard production. Redrawn in modified form by S.I.U. Cartographic Laboratory

emulsifying agents in current use tend to interfere with the sizing properties of the wax. Thus, it is generally attempted to keep the quantity
of the emulsifying agent to a minimum to obtain good sizing properties
from the wax used. Emulsifying agents, once subjected to heat and pressure in the hot press, are capable of releasing the wax droplets to melt
and cover particle surfaces.

To obtain efficient sizing, the wax used should be as widely distributed
over the particle surface area as is possible. Blending techniques, as well
as hot pressing and hot stacking (if practiced) are the major factors
affecting sizing maximization. Pecina,[11] in attempting to clarify some of
these points, mixed the paraffin wax to be blended with the particle
furnish with anthracene. The latter compound, upon solidification,
produces a strongly visible, greenish fluorescence under irradiation at a
wavelength of 300–400 mμ (ultraviolet light). In this manner, the
presence of wax could be detected readily in the adhesive droplets.
Pecina, spraying the resin-wax system containing anthracene, noted that
paraffin wax, used in the usual ranges in particleboard manufacture, was
found present in practically all adhesive droplets. The size of the emulsified wax particles are such that they can readily find room in the adhesive
droplets. In his measurements, Pecina noted that the adhesive droplets
ranged from 0.2 to 0.3 mm. in size while the wax particles measured only
from 0.004 to 0.014 mm. The latter particles are normally distributed
irregularly and in different amounts in the adhesive droplets similar to
that shown in figures 7–5A and 7–5B. Pecina, based on his experiments,
postulates that temperature alone is not sufficient for the wax particles
to impart water repellency. Such particles remain entrapped in the adhesive droplet while the latter solidifies under the influence of heat
(fig. 7–5C). The same can be said for the pressure alone: it breaks the wax
particles into yet smaller sizes which remain entrapped in the adhesive
(figs. 7–5D and 7–5E). On the other hand, a combination of pressure
and temperature not only breaks up the wax particles but also melts them,
thereby enabling them to combine to form a wax film (fig. 7–5F). This
phenomenon imparts water repellency to the wood particle surfaces on
which it occurs. From these observations, Pecina concludes that when
higher pressures and normal operating temperatures are used in hot
pressing, a greater distribution of wax occurs. Other variables being equal,
lower density particleboard as well as that made up of thicker and
coarser particles (resulting in larger inter-particle voids) will show a
greater tendency for liquid water absorption. This is due, in part, to the
fact that a large proportion of the wax-laden adhesive droplets is subjected only to proper temperatures in the hot press without encountering
adequate pressures.

7–5. *(A) Occluded paraffin wax in an adhesive droplet, (B) cross-sectional view of an adhesive droplet, (C) paraffin wax deposited in the cavities of an adhesive droplet when subjected to 125°C., (D) a single adhesive droplet, (E) same droplet when subjected to pressure only, and (F) combined effect of pressure and temperature on the droplet.*

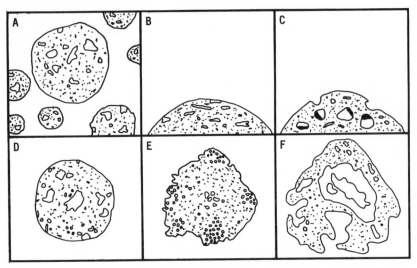

SOURCE: Pecina, H. 1965. Demonstration and action of the paraffin wax for water repellent processing of particleboard.
Redrawn in modified form by S.I.U. Cartographic Laboratory

7–3. Fire Retardants

Fire retardants consist of single or combinations of chemical compounds which are incorporated into particleboard to achieve a degree of fire retardancy. These chemicals can be added either during the course of manufacture or later impregnated into the product itself. The fire retardancy of these compounds may affect smoke emission, glow combustion, and flame spread, although many known compounds cannot retard all three at the same time. Fire retardancy imparted by these compounds is believed to be due to the changes they cause in pyrolysis (thermal degradation) reactions to charcoal formation resulting in reductions in the production of flammable components.[12] The reason for many retardant chemicals to have one effect on flame spread and another on smoke emission and glow is due to the fact that in wood pyrolysis the process of charcoal combustion is essentially independent of that responsible for flaming combustion of gases and liquid phases. It is not, therefore, surprising to find that of all efficient flame retardants only a very small number are also glow retardants; in fact, some may even influence glow in an adverse manner. Ideal fire retardants for particleboard must not only be compatible with raw material and process peculiarities, but they should also provide glow and smoke emission retardancies and stop flames from spreading. In many fires, the smoke produced often constitutes the major source of danger to life.

The use of fire retardants in particleboard is relatively recent. A greater degree of interest is being paid to fire retardants as the use of particleboard products in the field of housing and other types of construction gains momentum. In fire-retardant particleboard production, much use is made of the knowledge already available for solid wood, plywood, and textiles. However, in particleboard production, particular problems associated with raw material and manufacture are encountered.

Table 7–1 presents a list of basic compounds used either in research or production phases of particleboard and found effective in varying degrees when used alone or in combination. Of those listed in this table, ammonium phosphates and boric acid retard glow combustion in addition to the protection they offer against flame spread. Other flame retardants, e.g., ammonium sulfate, disodium octaborate tetrahydrate, and sodium arsenate, among others, actually provide for a more intense glow combustion. European sources using ammonium bromide note that, although this compound imparts a fair degree of fire retardancy and can be worked readily into the particleboard process, the resulting board could not pass the more rigid German specifications.[13] Boric acid has been found to impart better retardancy than ammonium bromide as well as to enhance strength and dimensional stability in laboratory-made boards. But these gains remain to be proven under production conditions. Plant experiences with boric acid in Belgium and Japan have not, so far, confirmed the laboratory results: six-foot-long panels retain a higher

7–1. *A partial list of basic fire retardant chemicals of varying effectiveness for use in particleboard*

Ammonium bromide	Zinc chloride
Boric acid	Zinc borate
Ammonium phosphates	Orthophosphoric acid
Chlorinated naphthalene	Dicyandiamide
Ammonium sulphate	Diammonium orthophosphate
Disodium octaborate tetrahydrate	Poliammonium orthophosphate
Sodium hydrogen phosphate	Sodium borate (anhydrous)
Diammonium phosphate	Sodium bichromate
Monoammonium phosphate	Phosphoric acid
Ammonium sulfamate	Sodium arsenate
Ammonium dihydrogen orthophosphate	Ammonium borates

level of moisture compared to controls. Using powdered boric acid in these instances has resulted in sifting and the accompanying uneven protection. European experience indicates that boric acid is to be preferred at present over many compounds examined including ammonium sulphate and acid phosphates for urea-bonded boards. This is due to the general belief that boric acid confers good fire retardancy without significantly degrading other desirable particleboard properties.

Johnson [14] utilized boric acid in both powder and solution forms in various proportions to determine its effectiveness in fire retardancy for laboratory-made particleboard. In his tests, boric acid-treated boards were compared to controls and those possessing monoammonium phosphate. Assuming the calculated fire rating of untreated red oak—in a modified Schlyter test—as 100, the results obtained showed that 10 percent or more boric acid can confer a considerable degree of fire retardancy (table 7–2).

Syska,[15] through a series of preliminary laboratory studies in which a number of fire retardant formulations were used, concluded that boric acid–disodium octaborate was the most satisfactory, among those tried, for particleboard. In these studies, the boric acid–disodium octaborate formulations were in the following proportions: 0:100, 10:90, 20:80,

7–2. Typical modified Schlyter test values of chemically treated particleboard

	Schlyter Test[1]		Calculated Flame Spread Rating
Treatment	Flamespread[2] (inches)	Time[3] (min)	
Red Oak	43	5	100
Untreated particleboard	43	3	165
5 percent Monoammonium Phosphate $(NH_4H_2PO_4)$ crystals	43	4.3	115
15 percent Monoammonium Phosphate $(NH_4H_2PO_4)$ crystals	13	——	30
7 percent Boric Acid (H_3BO_3) powder	25	——	58
15 percent Boric Acid (H_3BO_3) powder	11	——	26
10 percent Boric Acid (H_3BO_3) solution	11	——	26

[1] Tested for 10 minutes, 2500 Btu per cu ft at 4 cu ft per hr.
[2] Maximum height of flame above original flame height.
[3] Time to reach maximum flame height.

SOURCE: Johnson, W. P. 1964. Flame-retardant particleboard.

and 25:75. These were prepared at a solution concentration of 30 percent in water, heated to and maintained at 145° to 165°F. The disodium octaborate tetrahydrate was first dissolved, followed by the addition of the predetermined amount of boric acid. Syska incorporated, prior to drying, 10 to 15 percent of the fire retardant formulation (based on the oven-dry weight of the wood furnish) into particles. The treated particles were then dried, and boards were made in the usual manner. Tests revealed that, although there were practically no reductions in internal bond strength and the modulus of elasticity values compared to non-treated controls, modulus of rupture values had suffered by 25 percent. In another study, Arsenault[16] examined four fire retardant formulations in connection with producing fire-retardant aspen particleboard. Two of these formulations contained boric acid, although minor in proportions (see table 7–3). Various retention levels were incorporated into labora-tory-made boards in order to determine the levels at which effective fire retardancy is achieved. Figure 7–6 shows the influence of the various formulations and their retention levels on the maximum weight loss when subjected to the fire tube test (ASTM E69–50). It is noted that as the retention levels approached 5–6 pcf, sharp reductions in weight loss resulted with the exception of zinc borate (commonly used in fire-retar-dant paint formulations) which performed poorly in this respect. With the three more effective formulations, the temperatures developed over time during the fire test were also sharply reduced at retention levels of about 5–6 pcf. Orthophosphoric acid-dicyandiamide did not adversely affect the particleboard's hygroscopicity. This formulation, at retention levels of about 6 pcf, also appeared to confer good glow retardancy while imparting less color on the boards. Recent research has sought to incor-porate fire retardants into consolidated particleboard followed by a second hot pressing. In this technique, the fire retardant chemical is forced into the surface of the panel by hot pressing. Using ammonium dihydrogen orthophosphate and liquid ammonium polyphosphate, Shen and Fung[17] report that significant reductions in flame spread can be obtained.

The influence of fire retardant compounds on the resin binder is logically of utmost importance in particleboard manufacture. Practically all retardants examined so far result in varying degrees of strength deterioration, with some chemicals causing almost total strength loss. Furthermore, a great number adversely affect product hygroscopicity. The primary factor having deleterious influence upon the resin binder may be the lowering of the pH of the resin system by either the retardant chemicals or their thermal decomposition products. It should, therefore, be possible to avoid strength losses through proper pH control of the

7–3. Fire retardant formulations used by Arsenault

Formulation 1 (Minalith)		Formulation 2 (Pyresote)		Formulation 3	Formulation 4
Compound	*Percent*	*Compound*	*Percent*		
Diammonium orthophosphate	10	Zinc chloride	35	Zinc borate	An organic phosphate formed from orthophosphate acid and dicyandiamide in an equimolar solution D:P
Ammonium sulfate	60	Ammonium sulfate	35		
Sodium borate (anhydrous)	10	Boric Acid	25		
Boric acid	20	Sodium bichromate	5		

SOURCE: Arsenault, R. D. 1964. Fire retardant particleboard from treated flakes.

7–6. *The effect of various fire-retardant formulations on maximum weight loss of particleboard in a fire-tube test.*

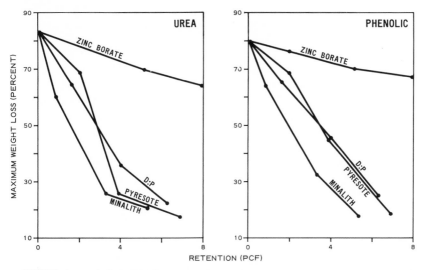

SOURCE: Arsenault, R. D. 1964. Fire retardant particleboard from treated flakes.
Redrawn in modified form by S.I.U. Cartographic Laboratory

resin and/or the fire retardants used. Nevertheless, it remains a prime requisite that fire retardants truly suitable for use in particleboard production must not, in any significant way, interfere with adhesion or adversely affect board hygroscopicity. Additionally, a qualified retardant chemical should be capable of utilization in existing particleboard production without any major modifications. Finally, it should be low in cost and compatible with product use. An increasing concern with fire safety has resulted in the development of modern building codes setting specific requirements for fire performance of construction materials. With ever-greater quantities of particleboard products being used in building construction, the rise in demand for fire-retardant-treated boards appears to be a strong possibility. Currently, there is a modest demand for fire-retardant-treated particleboard as core material with veneer overlays.

In summary, reasonably acceptable fire-retardant particleboard can currently be produced, however, at a cost both in terms of money and reductions in desirable board properties. Present fire retardants are not ideal for particleboard use. Some discolor the treated products and result in a higher degree of hygroscopicity. Many fire retardants known at present cause corrosion in metals and, when mixed in dry form with the particle furnish, can settle in air conveyors. On the positive side, some confer a degree of preservative protection on the board against biological attack.

7–4. Fungi and Insect Repellents

When particleboard is exposed to continuous warm and moist conditions over a relatively long period of time, it is subjected to fungal attack. Generally, particleboard not treated with chemical repellent is considered as resistant to attack by fungi as the species of wood or other lignocellulosic material which makes up the bulk of the board. This is also probably true relative to attacks by termites. European experience shows that particleboard is readily attacked by fungi if boards are moistened with water. Basidiomycetes such as some species of *Coniophora*, *Poria*, *Lenzites*, and *Polystictus* have been noted to attack particleboard in laboratory tests. Particle geometry has no influence on the severity of such attacks. Willeitner,[18] in conducting this study, further noted that particleboard bonded with urea-formaldehyde appeared to promote fungal growth while others bonded with phenol-formaldehyde tended to impede such growth. This is not to imply that phenolic-bonded boards are protected. Outdoor exposure studies have noted attack by fungi after just one year.[19] Both phenolic- and urea-bonded boards are susceptible to attack by fungi. Therefore, in applications where long-term and continuous high moisture content and temperature are to be encountered, effective preservative protection for boards appears to be necessary. When wood particleboard is used as floor underlayment over crawl spaces or slabs as well as in many outdoor applications, attack by biological agents cannot be discounted. If not protected, particleboard intended for use in tropical climates is also a candidate for attack by fungi (and termites). In fact, in hot and humid climates, boards can be attacked even during shipping and storage.

Insects which attack wood also attack particleboard. The most serious threat is posed by the various species of termites in the United States and other parts of the world where these insects are found. Conditions and building techniques which encourage attack by termites to wood are discussed elsewhere[20] and need not be repeated here. Generally, many conditions promoting fungal growth are also conducive to termite attack. The most effective additive for protecting particleboard against fungi and insects, including termites, appears to be pentachlorophenol or its sodium salt form, sodium pentachlorophenate. Although no absolute protection is guaranteed by the small quantities incorporated into particleboard, a fair degree of protection can be obtained.

Pentachlorophenol or sodium pentachlorophenate in the 0.25 to 2 percent range is added to the particleboard in an in-process step.[21] Either of these two preservative compounds can be incorporated into particle-board during the course of manufacture. From the standpoint of handling and satisfactory solubility, sodium pentachlorophenate is to be preferred over pentachlorophenol. It is available in powder, pellet, or briquette, and its technical grade contains approximately 83 percent active pentachlorophenol. It has been found more convenient to incorporate the required amount of sodium pentachlorophenate into the resin-sizing system to be sprayed onto the particles during blending. There is a disadvantage: sodium pentachlorophenate raises the pH of the resin system, thus forcing longer press times. This disadvantage may be overcome by neutralizing the preservative with acids. In the production lines where phenolic resins are utilized, sodium pentachlorophenate dissolves very readily due to the original alkalinity of such resins. Some believe that adding aqueous sodium pentachlorophenate to the undried particle furnish is to be preferred because it promotes crystallization of the preservative just below the wood particle surface leaving such surfaces relatively clean. [22] However, this technique may bring about the problem of particle break-age during normal handling of the furnish.

Adequate research data are not available on the influence of penta-chlorophenol on the strength and dimensional properties of particleboard. Some believe that the small quantities of pentachlorophenol added to particleboard do not adversely influence the strength and dimensional properties. Others show substantial drops in modulus of rupture when 1.25 percent (based on preservative solids and oven-dry furnish weight) is added to urea- and phenolic-bonded boards. [23] Both pentachlorophenol and sodium pentachlorophenate, under certain conditions, can leach out of particleboard. These compounds are highly toxic, and the products must not be used in applications where they endanger life. A deepening of color may be encountered in products containing these chemicals.

To conclude, the search for finding the type and quantities of pre-servatives to protect effectively all types of particleboard over long periods of time needs to be continued. The ideal preservative (as is the case for fire retardants) must not adversely affect desirable product properties and must be compatible with particleboard manufacture. With the projected increase in the variety of particleboard applications, particularly to out-door situations, a greater attention is likely to be placed on products truly protected against all biological agents.

7–5. Hardeners and Buffers

Hardeners and buffers are essentially chemical additives incorporated into the resin system usually just prior to the blending operation in order to make binders more responsive to the requirements of particleboard manufacture. Hardeners and buffers have already been discussed in conjunction with the resin binders, in chapter 6 (secs. 6–1 and 6–5).

Notes

1. Albrecht, J. W. 1968. The use of wax emulsions in particleboard production.
   Liiri, O. 1969. The most recent developments in the wood particleboard line.
2. Heebink, B. G. 1967. Wax in particleboards.
3. Heebink, B. G. 1967. Wax in particleboards.
   Kehr, E. 1964. Investigation of the suitability of various types and categories of wood for the manufacture of particleboard. Parts 3 and 4.
   Liiri, O. 1969. The most recent developments in the wood particleboard line.
4. Liiri, O. 1969. The most recent developments in the wood particleboard line.
5. Heebink, B. G. 1967. Wax in particleboards.
6. Albrecht, J. W. 1968. The use of wax emulsions in particleboard production.
7. Heebink, B. G. 1967. Wax in particleboards.
8. Stegmann, G., and Durst, J. 1964. Beech pressboard.
9. Not all researchers agree with these results. Some, in fact, contend that wax additions in the ranges discussed and particularly if held to within 0.5 percent, result in *increases* in strength under certain circumstances. (See Raddin, H. A. 1967. High frequency pressing and medium density board.)
10. Berrong, H. B. 1968. Efficiency in sizing of particleboard.
11. Pecina, H. 1966. Demonstration and action of the paraffin-glue for water repellent processing of particleboard.
12. Draganov, S. M. 1968. Fire retardants in particleboard.
13. Laidlow, R. A., and Hudson, R. W. 1969. Tour of chipboard factories and research institutes in Europe with special reference to the production of exterior quality boards.
14. Johnson, W. P. 1964. Flame-retardant particleboard.
15. Syska, A. D. 1969. Exploratory investigation of fire-retardant treatments for particleboard.
16. Arsenault, R. D. 1964. Fire retardant particleboard from treated flakes.
17. Shen, K. C., and Fung, D. P. C. 1972. A new method for making particleboard fire-retardant.
18. Willeitner, H. 1965. The behavior of wood particleboards under attack of basidiomycetes. Part I: Decomposition of particleboards by basidiomycetes.
19. Gressel, P. 1969. Investigations on weather aging of particleboard: A comparison between two-year weathering and three different accelerated test methods.
20. Hunt, G. M., and Garrett, G. A. 1967. Wood preservation.
21. A European source recommends that, in external situations, particleboard should also be protected with ex-process treatments such as, coating the products with a dense polyurethane paint or polyvinyl chloride sheeting. (See Laidlow, R. A., and Hudson, R. W. 1969. Tour of chipboard factories and research institutes in Europe with special reference to the production of exterior quality boards.)
22. Huber, H. A. 1958. Preservation of particleboards and hardboard with pentachlorophenol.
23. Brown, G. E., and Alden, H. M. 1960. Protection from termites: Penta for particleboard.

Density, Layering, and Other Boards

8–1. Density and Use

Product density in particleboard is of prime importance due to the influence it exerts on physical, mechanical, and industrial characteristics of the resultant product. Higher board densities are associated with higher strengths, more difficult machining characteristics, a greater degree of dimensional instability, and higher cost per unit volume. On the contrary, low-density board offers better insulating characteristics, higher dimensional stability, lower strength, and less unit cost. In addition, the direct influence of density over product weight is of basic importance in many applications.

The density of particleboard may vary over a wide range—depending on the technological practice and raw material makeup. Particleboard is produced in densities ranging from 0.25 gm/cm³ (15 pcf) to 1.20 gm/cm³ (75 pcf). Most particleboards currently produced possess moderate densities ranging from 0.40 gm/cm³ (25 pcf) to 0.80 gm/cm³ (50 pcf). When speaking of density, particleboard is frequently divided into three broad categories: low density, which is less than 37 pcf, medium density (37–50 pcf), and high density with a value over 50 pcf. Commercial standard CS236–66, using this density categorization, divides particleboards into ten different classes based on their minimum strength and dimensional characteristics. Specialty-type particleboard such as "Flapreg,"[1] forms an example in which higher densities in the range of 1.0 gm/cm³ or more are involved (see section 8–4).

In making high-density particleboard, sometimes fibers are flat-pressed into smaller thicknesses ($\frac{3}{8}$ inch or less). This makes such boards resemble hardboard and, in many uses, can be substituted for the latter. However, the basic difference between high-density particleboard and hardboard lies in the means utilized to bond the lignocellulosic particles: in most standard hardboard, major reliance is placed on what is believed to be the natural bonding action of depolymerized lignin, while in particleboard synthetic resins form the binder. The use of smaller particle sizes is not a necessary requirement in the manufacture of high-density particleboard. Larger particle sizes, particularly flakes can also readily be used.

Low-density particleboard has limited production and is usually used in applications where insulation and weight considerations are important.[2] It can be used as panel material where both heat and sound insulation are required, or as core material for veneered and sandwich structures. The use of low-density particleboard as core material has been specially limited due to a number of shortcomings, such as low resistance to withdrawal of screws or other mechanical fasteners. The strength of low-density boards is low but can be engineered in such a way that adequate strength for the intended application is obtained. Thickness of low-density particleboard has been limited to approximately a maximum of one inch when conventional platen heating is employed. With high-frequency heating facilities, thicker boards may be economically produced.

A relatively recent innovation in the low-density particleboard field has been the development of thick panels in which both very coarse chips and flake-type particles are employed. A three-layer structure utilizes pulp chips for the core layer, with flakes forming the surfaces. It is produced in thicknesses equaling or exceeding $1\frac{3}{4}$ inches and is intended as possible roof decking material. The product, developed at the United States Forest Products Laboratory, Madison, Wisconsin, and refered to earlier has a density range of 22–25 pcf and appears to meet the roof decking requirements with a span of 48 inches.[3] Preliminary tests have indicated that pulp chips used should be of low-density species in order to keep the overall panel density in the low range. Some flexibility is permitted as to the species utilized for flakes.

Medium-density particleboard has fulfilled the requirements of more end uses than other density groups. Most particleboard currently being produced can be categorized as medium density. Both single and multi-layer boards in panel sizes of 4- by 8-foot or larger are extensively produced. Moderate strength and dimensional stability are obtained for this type of particleboard. The density of medium-density particleboard can average from zero to 100 percent higher than the species of wood or other lignocellulosic material from which it is made. In most cases, however, the board has a density 25 to 50 percent higher.

The use of medium- and low-density particleboard as core material for furniture panels, flush doors, and veneer wall panels is common. These uses will probably always be of major importance due to inherent advantages of particleboard for such applications. The availability of large, smooth, and jointless panels that are readily worked has been a significant advantage to many users of particleboard. Due to lower cost and the fact that some particleboard properties are superior to lumber, particleboard has been increasingly preferred over lumber for core material. High quality furniture is being produced with a large degree of gloss and no danger of sunken joints. Furniture components made with particleboard may be overlayed with veneer, plastics, or hardboard. Edges and ends of panels may be banded with solid wood. Research and development in recent years has made edge banding unnecessary through the use of various edge treatments and by improving the structural characteristics of the board itself. As core stock, medium-density particleboard is used for desk tops, drawer fronts, chest and table tops, dust dividers, shelves, chest sides, sewing machine cabinets, bed rails and headboards, bookcase sides, backs and shelves. Display fixtures and doors, restaurant furniture and bars, and storage fixtures use particleboard cores. A major problem in earlier years, in connection with the use of particleboard, was the possibility of telegraphing through overlaid panels on which high gloss finishes were applied. To avoid this problem, crossbanding with veneers was employed. With the development of highly smooth surfaces through the use of a number of technological and raw material innovations, such telegraphing has been eliminated making the veneer crossbanding unnecessary.

The use of medium-density particleboard for floor underlayment has been, and continues to be, of major proportion. In addition to floor underlayment, particleboard is being employed for interior ceilings and walls, components for interior millwork such as kitchen cabinets, wardrobe, and other storage compartments. In Europe, in addition to furniture, interior furnishing and construction, particleboard has increasingly gained significance in various special fields, e.g., ship interiors, vehicle construction, and consoles for electronic equipment. Other uses in mobile home construction and advertising displays are gaining economic importance rapidly. Recent research in the United States on the development of permanent type low-cost housing, holds the possibility of substantial uses for particleboard, some of which is designed to support structural loads.

A significant development in the particleboard field over the last decade has been the production of exterior panels which are intended for service in semiexposed or exposed outdoor conditions. Heebink[4] classifies exterior particleboard of medium density into two general types depending on use requirements. The first type has at least one smooth side and is primarily intended for nonstructural applications. Such exterior boards may be used in house siding, soffits, porch ceilings, boat panels, highway signs, and cargo trailer floors and partitions. Strength would be of minor importance in these uses. However, dimensional stability would be very important. The second type is required to meet certain structural requirements, but its surface appearance would be of little consequence. Examples of the use for the second type include wall and roof sheathing, pallets, shipping containers, rough structural panels, and sheathing for crates. In these applications, the product is often used in situations where it is fastened by nails; thus nail withdrawal resistance must be adequately high. In addition, such applications require that the board be strong enough to supply the structure with proper rigidity and stiffness and in some cases with impact resistance. Be it the first or second type, the board is required to withstand the rigors of its service environment without serious degradation. Degradation occurs in one or more of a variety of ways such as excessive thickness swelling, surface roughening, linear movement, strength reduction, or product distortion. When objectionable degradations occur, usually more than one of the above characteristics emerge due to the interrelationships involved. For instance, thickness swelling or surface roughening may accompany strength reduction.

8-2. Layering

Particleboard is made in either randomly distributed single layers or in a number of layers. Layering is essentially a lamination process in which the laminates are discontinuous regardless of particle geometry and the technological practice. In the lamination process relevant to particleboard manufacture, adhesive spreads are very low compared to those in laminating lumber or veneer.

In particleboard structures made up of more than a single layer, a basic characteristic involves the differential strength properties between the face and core layers. This type of differentiation permits densification at surface layers designed to withstand higher bending moments without undue imposition on cost and product weight. For instance in a three-layer board, the faces may have a density of g_f and the core g_c in which:

$$g_f > g_c.$$

The resultant density of the entire board, g_R would then be:

$$g_R = \frac{g_c(t-2t_f) + 2g_f t_f}{t} \tag{8-1}$$

in which t and t_f refer to the thickness of the entire board and that of each face respectively. In terms of the shelling ratio:

$$\lambda = \frac{2t_f}{t}.$$

Equation 8–1 can be rewritten as:

$$g_R = g_c + \lambda(g_f - g_c). \tag{8-2}$$

To obtain the strength differentiation, layering permits the use of a number of techniques such as:

(a) A higher adhesive content in the face layers.
(b) Different particles (e.g., smaller, thinner) in the face layers.
(c) Lower density wood species in the face layers.
(d) Processing techniques which reduce compression strength of the face particles, i.e., using particles of higher moisture content.
(f) Surface particle orientation to achieve higher bending strength in the direction of orientation with other parameters (adhesive content, particle size, etc.) kept constant.

Upon close examination of the first four techniques just outlined, it is noted that they are aimed at increasing particle-particle contact. In connection with the last technique, it is worth noting that particleboards have been developed[5] in which the core consists of veneer with the faces made up of resin-bonded flakes oriented perpendicular to the grain direction of the veneer. Tests have shown that bending strength similar to that of plywood can be obtained in panel construction of this type. The strength of the assembly may be varied through manipulation of the density and resin content of the particle faces.

8-3. Foamed Particleboard

Early trials using polyurethane foam as the binder show promise in the production of lightweight, dimensionally stable particleboard with adequate strength characteristics.

The production of polyurethane in the United States dates back to the mid-1950s, with prior work in development and production carried on in Germany. Polyurethane foams are basically the result of a reaction between an isocyanate (R–NCO) and a compound containing active hydrogen atoms such as a polyester or polyether. The additional reaction triggers changes which lead to polymerization with the result being the production of a rigid plastic. The reaction, in its simple form, can be stated as follows:

$$\underset{\text{polyester}}{R_1 - C\overset{O}{\underset{OH}{}}} + \underset{\text{isocyanate}}{RN = C = O} \rightarrow \underset{\text{long-chain polyurethane}}{R_1 - C\overset{O}{\underset{\underset{R - N - C\overset{\backslash}{\underset{O}{}}}{\overset{|}{H}}}{O}}}$$

$$\overset{R_1 - C\overset{O}{\underset{\underset{R - N - C\overset{}{\underset{O}{}}}{\overset{|}{H}}}{O}}}{} \xrightarrow[\text{heat}]{A} R_1 - \overset{O}{\overset{\|}{C}} - \overset{H}{\overset{|}{N}} - R + CO_2$$

polyurethane

The reactions above show a polymerization reaction between the hydroxyl groups of the liquid polyester and the isocyanate to form long-chain, high molecular weight urethane polymers. The degree of reactivity of the isocyanate and the hydroxyl containing compound may be intensified through the use of catalysts such as metal compounds, tertiary amines or alkaline-reacting chemicals. Currently, an aromatic polyisocyanate such as diisocyanate is combined with glycol hydroxyl groups in high molecular polyesters along with accelerators to produce polyurethane foam plastics. Agents consisting of low blowing substances such as trichlorofluoromethane are often used to assist foaming. The more the number of functional groups, namely −OH and −NCO in polyester and diisocyanate respectively, the higher the possibility of cross-linking and the more rigid the resulting plastics. Also relevant to plastic rigidity is the ratio of the isocyanate used to the polyester. In summary, the production of polyurethane foam can be described as follows:

(poly) isocyanate + polyester + accelerator + blowing agent
= polyurethane foam

In production of polyurethane foam, the probability exists that the wood itself may participate in the reaction by providing hydroxyl groups present in cellulose chains:[6]

The $-OH$ groups probably react with $-NCO$ of the isocyanate to advance the development of the polyurethane. However, an insufficient number of such $-OH$ groups would normally be present in various particle configurations employed in particleboard to eliminate the need for polyesters.

In production of particleboard with polyurethane foams, the particularities relative to the use of such foams must be considered and the process accordingly designed. When the activator is added, foaming and subsequent hardening will occur in the matter of a few minutes. Table 8–1 illustrates that, when there is a 0.5 percent activator addition (based on liquid plastic weight), foaming is initiated in thirty seconds with the plastic assuming a rigid state in two minutes. Whatever process is employed, the pressing therefore must follow the urethane addition in very short order. The procedure of blending resin with the furnish, followed when conventional resins are used, would not be suitable with polyurethane. The heat generated in the blender along with the time factor may

8–1. *Starting and curing times for various activator addition levels in polyurethane foams*

Activator Addition[1] (percent)	Starting Time[2] (sec)	Curing Time (min)
0.0	120	20
0.5	30	2
1.0	15	1

[1] based on liquid plastic weight
[2] time needed to initiate foaming

SOURCE: Deppe, H. J. 1969. Developments in the production of multi-layer foamed wood particleboard.

8–1. *Schematic layout of particleboard production using polyurethane foam with (A) intermittent single daylight press, and (B) continuous press.*

A. INTERMITTENT

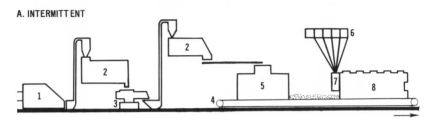

B. CONTINUOUS

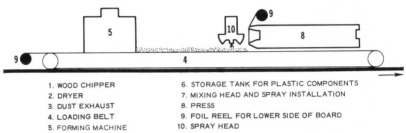

1. WOOD CHIPPER	6. STORAGE TANK FOR PLASTIC COMPONENTS
2. DRYER	7. MIXING HEAD AND SPRAY INSTALLATION
3. DUST EXHAUST	8. PRESS
4. LOADING BELT	9. FOIL REEL FOR LOWER SIDE OF BOARD
5. FORMING MACHINE	10. SPRAY HEAD

SOURCE: Deppe, H. J. 1969. Developments in the production of multi-layer foamed wood particleboard.
Redrawn in modified form by S.I.U. Cartographic Laboratory

result in loss of blowing agent and premature foaming. Low-boiling compounds used in polyurethane foam formulations require very small quantities of heat for activation. This characteristic, however, can be turned into an advantage by eliminating the conventional blender in the process and incorporating the plastic material into the already-formed mat shortly before it undergoes pressing. Deppe,[7] utilizing existing particleboard production lines proposes two setups, one on an intermittent basis and the other continuous (fig. 8–1). In either setup, furnish preparation is undertaken as usual, with minor exceptions: dustlike particles should be removed from the furnish since they are not readily surrounded by the foam. In the polyurethane preparation stage, separate chemical components are combined, one batch at a time, in a mixing chamber which is subsequently discharged on the already-formed mat. The liquid plastic may be discharged employing various techniques, e.g., mobile dispensing head traveling across the mat or a device similar to a curtain coater. Once the liquid plastic is dispensed onto the mat, subsequent foaming action drives the material in and around the particles through the mat before it attains rigidity in the cold press. Minor problems associated with this intermittent technique include cleaning of deposits in the mixing chamber which may build up and trigger undesirable reactions in subsequent batches. A number of techniques using recoverable solvents or diisocyanate (which is subsequently dispensed onto the mat) have been used. This is not, however, a problem in continuous operations (fig. 8–1), since regular intermittent interruptions, such as that occurring in noncontinuous setups, would not occur.

Polyurethane, too, will tend to adhere to cauls or other metal surfaces coming into contact with the mat during pressing. This can be eliminated by various techniques such as flame siliconization.[8] Also, when the liquid plastic is dispensed over the formed mat, foaming may cause a portion of the material to reach over the edges and onto the caul plate or the metal belt. To eliminate this problem, the plastic material should not be dispensed over a distance of approximately an inch from the edges. These edges are subsequently returned to the production line. Both intermittent and continuous operations utilize single-opening presses due to the relatively rapid onset of foaming and the need for quick pressing. Overlays, i.e., veneer, plywood, foil, or fiber glass mats, are placed on the particle-polyurethane mat and pressed simultaneously.

Polyurethane-bonded particleboard ($\frac{2}{3}$ of weight being particles and about $\frac{1}{3}$ polyurethane) possesses properties which make it attractive for use. At lower densities than equivalent resin-bonded board, it has reasonably good (although somewhat lower) bending strength but substantially higher internal bond strength. Superior nail- and screw-holding characteristics exceeding those of resin-bonded particleboard can be obtained. Perhaps the most attractive property of particleboard made with polyurethane is its dimensional stability. The board in contact with liquid water absorbs an insignificant amount or no water at all even after long periods of soaking (six months) under European standard conditions.[9] Its bending strength can be significantly improved with the use of strong overlays such as veneer. Polyurethane particleboard, however, is susceptible to attack by fungi. Its fire retardancy may also be adversely affected. The addition of fire retardant chemicals may have to be considered in order to meet specification requirements dealing with board combustibility. The major drawback in the production of polyurethane particleboard at the present time is economic. The cost of the polyurethane required is prohibitively high when compared to conventional resins. According to one European study, polyurethane expenditure exceeds $\frac{2}{3}$ of the total raw material costs. To reduce such costs, the plastic material can be extended, but this leads to a board with less favorable properties.

Further research on processing techniques both in the field of foamed particleboard and plastics may make the production of this type of product economically competitive in the future.

8–4. Other Plastic- and Resin-Impregnated Products

Some innovations in the field of particleboard have involved incorporation of plastics or resins into the particles themselves or the resultant board. Various plastics (e.g., the acrylates, a styrol and vinyl derivative) in the monomeric form are used to impregnate the resultant board in much the same way as it is used for solid wood. In the last decade, appreciable knowledge has been gained with respect to what is often termed as a "wood plastic combination" in which wood of various species is impregnated with liquid monomeric plastics and subsequently polymerized *in situ*. The same principle and procedure are applicable with particleboard.[10] The polymerization of polymer is obtained by the use of radioactive radiation, heat treatment, or chemical compounds. Improvement and modification of properties achieved in solid wood can probably be realized with particleboard. Specialized products could therefore be manufactured which are appreciably different from ordinary particleboard and are capable of meeting an entirely new set of specifications.

Another product of older origin involves the application of "compreg" production to particleboard. In producing a plywood specialty product called compreg, a water solution of low molecular weight phenol-formaldehyde resin is allowed to diffuse into sheets of veneer. The sheets are then placed in an environment with moderate temperatures for sufficient lengths of time to allow the water to evaporate without significantly advancing the resin. Wood veneers so treated are then built up to the required number of plies and hot-pressed at high pressures. The product has considerably higher density (greater than 1) but it is characterized by its much higher strength and dimensional stability compared to ordinary plywood. The dimensional stabilization achieved by this technique is brought about by resin polymerization within the cell wall structure of the wood.

In "flapreg,"[11] flakes are used instead of sheets of veneer. The procedure follows a combination of those employed for compreg and particleboard production. In making flapreg, particles are produced and dried in much the same way as in particleboard production. Once the particle moisture content is down to approximately 7 percent or less, they are ready for resin impregnation. The required amount of resin solution is then sprayed on the furnish since immersing the flakes will cause clumping and create

8–2. *Properties of "flapreg"*

Property	Value[1]
Density (gm/cm³)	1.39
Modulus of rupture (psi)	13,170
Modulus of elasticity (psi)	1,910,000
Internal bond (psi)	1,097
Thickness swelling (percent) 24-hr. soak	0.42
Water absorption (percent) 24-hr. soak	0.44

[1] Data were obtained from boards made as follows:
 Flake size: l: $\frac{1}{2}$ in., w: $\frac{1}{4}$ in., t: 0.008 in.
 Resin content: 35 percent solids, phenol-formaldehyde (dry wood basis)
 Moisture at pressing: 7 percent (dry wood basis)
 Pressing: temp: 325°F, press: 2500 psi, time: 25 min., cooling time: 5 min.
 SOURCE: Talbott, J. W. 1959. Flapreg flakeboard—resin impregnated compressed flakes.

problems in mat formation. Once the resin is sprayed, the impregnated particles are allowed to redry at a moderate temperature to permit water evaporation without resin polymerization. After redrying, the particles are formed into a mat normally of much greater thickness than that encountered in production of medium-density particleboard. The mat, therefore, may have to be prepressed in order to keep the daylight openings in the hot press within a manageable dimension. In hot-pressing (at 300–350°F.) the mat requires longer press times (fifteen to twenty-five minutes) and higher pressures (700–2500 psi) compared to those required in production of particleboard, but a product of enhanced properties and higher density will emerge. The nature of raw materials and press cycle in flapreg production is such that blistering may occur upon press opening unless the furnish moisture is kept at a very low level; otherwise, the press may have to be cooled just before the pressure is released.

Table 8–2 shows improved strength and dimensional stability as well as greatly increased density compared to properties of particleboard. A variety of property specifications can be met by modifications in resin impregnation and density. Also, particles may be formed in layered or directional patterns in much the same way as it is possible with particleboard.

In working with flapreg, carbide tools or high speed steel are used. The product has a tough, smooth surface and needs no other finish for most applications. Sanding imparts a high degree of polish to the surface which is "cigarette-proof" at higher densities. It is highly resistant to organic solvents and acids but is slowly attacked by alkali. Flapreg can be produced in other than flat shapes provided that only a very limited amount of flow is required. Such products as trays, toilet seats, and those with other simple shapes may be pressed with flapreg. In flat form, flapreg has potential uses as electrical insulation, floor tile, table and counter tops, cutting boards, electrical parts, gears, cams, metal spinning chucks, patterns, punch and die sets for sheet metal forming, cutlery handles, saw handles, golf club heads, and brush handles.

Notes

1. An abbreviated combination of "flake" and "impregnation." (See Talbott, J. W. 1959. Flapreg flakeboard—resin impregnated compressed flakes.)

2. Anon. 1957. Fiberboard and particleboard. FAO, Rome.
   Oberlein, A. 1957. Particleboards as panel materials.
   Wilhelmi, H. 1957. The use of particleboards in building and construction work.

3. Heebink, B. G., and Lewis, W. C. 1967. Thick particleboards with pulp chip cores—possibilities as roof decking.

4. Heebink, B. G. 1967. A look at degradation in particleboards for exterior use.

5. Elmendorf, A. 1967. New tri-panel combines veneer core and wood strands in product that utilizes entire log.
   Patent (U. S.) 1965. Patent No. 3,202,743.

6. Schulman, J., and Wilner, B. L. 1960. New uses for wood flour in rigid polyurethane foams.

7. Deppe, H. J. 1969. Developments in the production of multi-layer foamed wood particleboard.

8. Commercial Standard CS236–66. 1966. Mat-formed wood particleboard.

9. Deppe, H. J. 1969. Developments in the production of multi-layer foamed wood particleboard.

10. Liiri, O. 1969. The most recent developments in the wood particleboard line.

11. Talbott, J. W. 1959. Flapreg flakeboard—resin impregnated compressed flakes.

Dimensional, Thermal, and Acoustical Properties

9-1. Thickness Movements

Particleboard is largely made of wood particles, a basic characteristic of which is hygroscopicity. The presence of moisture in the surrounding atmosphere therefore is related to the amount of water present in the wood. Hygroscopicity, in turn, is reflected by the particleboard.

In particleboard, dimensional changes perpendicular to the plane of the board (thickness movement) will largely reflect changes across the grain (radial and tangential directions). The wood particles themselves, too, as the result of pressing, are in a compressed state once they emerge from the press. The extent of this compression depends on the density to which the board is compressed compared to the density of the wood species from which it is made. In most wood species, the maximum across-the-grain movement may be limited to approximately 6 percent in the tangential direction based on oven-dry dimensions. In particleboard, however, thickness change may amount to 20 percent or more. The question, therefore, arises as to what is responsible for such large thickness movements besides that accountable to wood hygroscopicity. The answer lies, for the most part, in the compressed state of particles in the board. The expansion component due to the latter phenomenon is referred to as "springback" and is brought about by the release of residual stresses which keep the particles themselves compressed. The release mechanism is readily triggered by moisture gain in the panel.

Along with particle hygroscopicity and springback, water is also absorbed by the binder material. The increase in moisture causes deterioration in the bond and results in particle-particle separation. This will contribute to the overall thickness swelling in the panel. Particle-particle separation may also be brought about by density differences over the surface of the panel. A given panel has areas whose densities will be lower or higher than the average panel density. Once moisture absorption occurs, this density difference results in differential thickness swelling. The areas of higher density tend to swell more compared to areas of lower density. The localized high density areas will subsequently have a tendency to swell out the low density areas beyond the limit which they would normally reach were they detached. Swelling stresses are generated which are capable of inducing partial failures in the wood particle itself and the particle-particle bond.[1] Such factors as species, type and level of resin addition, particle geometry, type and amount of additives, furnish and board moisture content (upon emerging from the hot press), pressing parameters, and product density all affect the stability of the product.

When particleboard is first exposed to moisture after pressing (in either liquid or vapor forms), swelling is brought about as already discussed. However, upon redrying, it will be noted that only a portion of the swelling will be recovered. The portion which does not recover is the springback component (fig. 9–1). Additional amounts of springback will be added in successive absorption-desorption cycles. The magnitude

9–1. *A schematic drawing showing the recoverable and non-recoverable (springback) proportion of thickness swelling.*

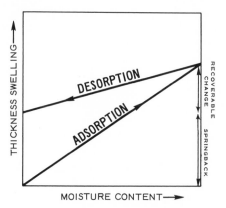

Drawing by S.I.U. Cartographic Laboratory

9-2. *Occurrence of permanent thickness swelling with cyclic exposure for boards of four different thicknesses. Upper depicts the results for three exposure cycles (each including 3 weeks at 95 percent followed by 3 weeks at 30 percent relative humidity) for indicated board thicknesses. The lower shows the results for five exposure cycles (each including 3 days of soaking followed by 11 days of drying).*

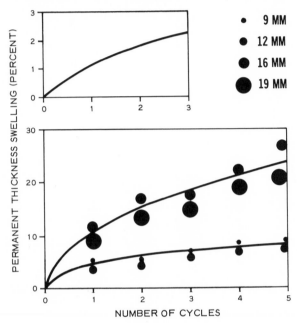

SOURCE: Liiri, O. 1961. Investigation of long-term humidification and the effect of alternate moisture variations on the properties of particleboards.
Redrawn in modified form by S.I.U. Cartographic Laboratory

of springback per cycle will be reduced with advancing cycles. Springback is encountered whether cycles include exposure to water vapor or liquid water (fig. 9–2).[2]

Springback is undesirable for various reasons. The first is the obvious effect on the permanent thickness increase with its deleterious influence on the thickness tolerance. In addition, the swelling process, in general, and the springback, in particular, bring about gradual structural deterioration resulting in lower strength[3] and degraded surface appearance. A reduction or elimination of springback would greatly enhance thickness stability, thereby favorably affecting the durability of the product.

Besides, the reversible thickness change would not have permanent damaging effect on the board structure. A number of avenues have been followed in attempts to reduce thickness swelling by either relaxation of built-in compressive forces in the panels or reduction of wood hygroscopicity (see also sec. 9–3).

In the development of an acceptable exterior particleboard, thickness stability is of utmost importance since the severe elements to which the product is subjected will encourage extreme movements in thickness. Should such movements be high, rapid deterioration both in mechanical properties and appearance will set in. Thickness swelling in exterior particleboard accompanies surface roughening (also referred to as fiber popping). Surface roughening, in reality, is a localized manifestation of swelling in which some particles rise above others. It may be brought about by rain, roof leaks, or vapor condensation in painted or overlaid panels located on the outside surface of a house. In cold climates, the use of particleboard as siding, particularly when vapor barriers are absent, constitutes a very severe service condition in which moisture movement, during the heating season, is likely to cause saturation of the boards. Unless such boards are relatively stable, significant degradation will likely result. Dimensional instability can also cause buckling and warping of panels and loosen fastening devices. Should thickness swelling be substantially reduced, a wide range of specifications for outdoor service conditions can be met by resin-bonded particleboard. Such uses include construction (both commercial and residential), marine, and signs applications.

The wood species utilized exerts a noticeable effect on thickness stability. Generally, particleboard made of lower density species, keeping panel density and all other factors constant, will exhibit a greater degree of thickness swelling. Figure 9–3 illustrates this point: single-layer boards made of pine (with no wax addition) show a greater degree of thickness swelling compared to those made of beech when soaked in water for two or twenty-four hours. The difference due to species remains over a wide range of panel densities. For example, at a panel density of 0.60 gm/cm^3, thickness swelling for particleboard made of pine in the study represented by figure 9–3, is on the average some 2.5 percent higher than that recorded for beech particleboard. The reason for this difference is believed to be due to a greater degree of mat compaction taking place when lower density wood species are used.

Particle geometry, too, influences thickness movement. When long, thin particles (flakes) are used, they essentially fall flat in the forming operation. This type of positioning imposes the greatest degree of hygroscopic movement in the panel thickness direction since movements

9–3. *Relationship of board density and species with thickness swelling.*

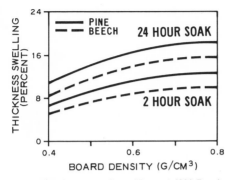

SOURCE: Stegmann, G., and Durst, J. 1964. Beech pressboard.
    Redrawn in modified form by S.I.U. Cartographic Laboratory

across the grain of particles will be reflected. On the other hand, this situation contributes to high linear stability. Shorter, thicker particles, however, produce panels of lower thickness movements but higher linear instability due to a larger proportion of such particles falling vertically during forming (fig. 9–4).

The type of binder has an insignificant effect on thickness stability under mild exposure conditions; however, it becomes critical when such conditions are severe. Urea-formaldehyde resins generally cannot withstand severe exposure, i.e., long-term high humidity and temperature or wetting. The particle-particle bond will deteriorate, resulting in a very large thickness swell. A considerable amount of research carried out on plywood over the last two-and-a-half decades, indicates increasing exposure resistance to thickness swelling and general bond degradation in the following order:

(a) Urea resins.
(b) Melamine-reinforced urea resins.
(c) Melamine resins.
(d) Phenolic resins.

A more limited research on particleboard indicates the same. The development of exterior particleboard has, therefore, been exclusively centered around the use of phenolic resins.

The amount of resin used is one of the most important factors of influence with regard to thickness movement. Greater additions of resin

9–4. *Influence of particle geometry, resin content, and panel density on thickness swelling as tested under laboratory and outdoor conditions.*

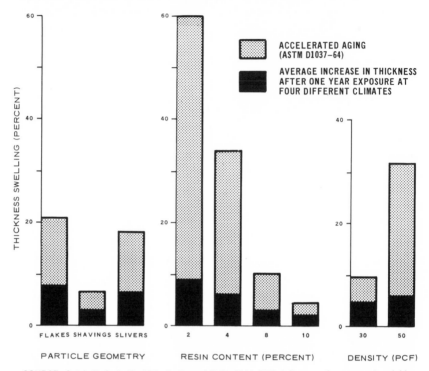

SOURCE: Gatchell, C. J., Heebink, B. G., and Hefty, F. V. 1966. Influence of component variables on properties of particleboard for exterior use.
Redrawn in modified form by S.I.U. Cartographic Laboratory

translate into a board of enhanced properties—particularly its thickness stability. Figure 9–4 illustrates how dramatically thickness swelling is affected as the resin content is increased from 2 to 10 percent (based on resin solids and dry weight of furnish). In figure 9–4, it is noted that thickness swelling is reduced from some 60 percent to less than 5 percent as the adhesive (phenol-formaldehyde) is increased over the indicated range. However, the level of adhesive that can be added must be considered carefully from the economic point of view due to the high costs of resin.

Higher density boards will swell more if subjected to water for a sufficient length of time. Figure 9–3 illustrates that as the panel density is increased over the 0.40–0.80 gm/cm³ range, high thickness swellings will result. Using ASTM Accelerated Aging tests (D1037–64) and phenolic-bonded Douglas-fir particleboard, the influence of panel density on thickness swelling is shown in figure 9–4. Lower density panels exhibit less swelling for two probable reasons. First, they contain less wood per unit volume with the ensuing swelling partially extended into interparticle voids, and, second, the existence of a smaller volume of wood per unit area which will mean a lower degree of hygroscopic response. Obviously, the overall panel thickness measured is the total thickness response of all the layers. Practically all types of particleboard, due to the nature of their processing, have a differential density over the thickness profile as we have already seen. This will result in a differential thickness swelling illustrated by figure 9–5 obtained for a single-layer phenolic-bonded commercial particleboard with an overall density of 0.768 gm/cm^3 and 4.2 percent adhesive content. The test conditions for arriving at this relationship consisted of equalizing at 65 percent relative humidity and 70°F. followed by transfer to 95–98 percent relative humidity and 80°F.

9–5. *Swelling data for half of thickness (divided into 13 layers) for a single-layer phenolic-bonded particleboard.*

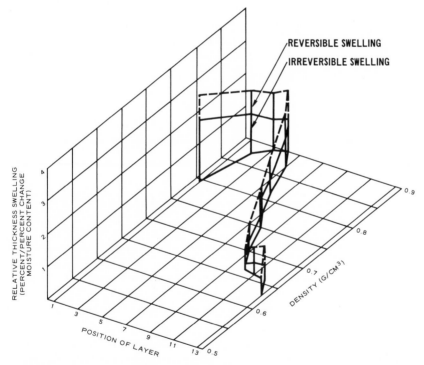

SOURCE: Suchsland, O. 1966. Some performance characteristics of "interior" and "exterior" type particleboard.
Redrawn in modified form by S.I.U. Cartographic Laboratory

Reconditioning the test specimens at 65 percent relative humidity and 70°F. completed the cycle. In this study, half of the board thickness was divided into thirteen thin layers and swelling for each was determined.[4]

In three-layer particleboard, the overall thickness change is the result of both surface and core layers. Thus, the resultant swelling S_r can be calculated as follows:[5]

$$S_r = S_c + \lambda(S_f - S_c) \qquad (9-1)$$

in which S_f and S_c signify the face and core layer swellings and λ is the shelling ratio. Equation 9-1 has shown close agreement with actual measurements. It indicates that should the face layers be extremely thin, the swelling behavior will approach that of a single-layer board. In equation 9-1 thickness change with moisture has been assumed linear, with S_r, S_c, and S_f representing specific swelling values (in percent per percent change in moisture content).

Other raw material and process variables, i.e., furnish moisture content, press time, and the panel moisture content (as it emerges from the hot press), appear to have varying degrees of influence on thickness movements. The results depend to a large extent, on other material and process parameters selected and the type of tests employed. Generally, longer press times and drier panels would mean a greater degree of thickness swelling. Higher furnish moisture appears to result in a less dimensionally stable board.

Testing conditions will logically have an influence on the dimensional stability data obtained. In some tests, the specimens are subjected to high humidity environments; in others, they are soaked in water. While the former is more appropriate for products intended for interior service conditions, the latter would be applicable to those destined for outdoors. When specimens are soaked in water, the duration of soaking, water temperature, and specimen size have appreciable influence on the data gathered. The longer the samples are soaked, the greater will be the amount of water absorbed. Longer soak times will adversely influence the particle-particle bond to a greater extent. Figure 9-6 shows the triple effects of water temperature, specimen size, and duration of soaking on thickness swelling for a three-layer, one-inch-thick particleboard. It is noted from this figure that substantially different results are obtained when the water temperature varies from one test to another. Specimen size, too, has significant influence on data gathered. Smaller specimen sizes would mean larger swelling if the soaking period is relatively short (two or twenty-four hours). However, when soaking periods becomes longer, the specimen size differences become irrelevant. Smaller specimens attain a more thoroughly saturated state during short intervals of soaking

9–6. *Effect of water temperature, specimen size and duration of soaking on thickness swelling.*

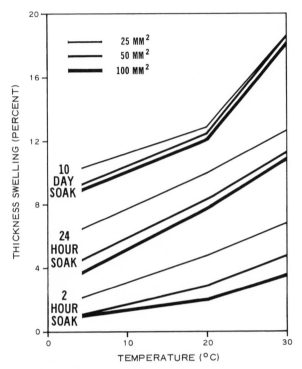

SOURCE: Neusser, H., Krames, U., and Haidinger, K. 1965. The response of particleboard to moisture with special regard to swelling. Redrawn in modified form by S.I.U. Cartographic Laboratory

with the larger specimens gradually reaching the same state as the soak time is extended. In all cases, water penetration in submerged specimens takes place mostly through the edges. As the specimen size becomes larger, thickness swelling at the central sections of the specimen tend to lag behind the edges. Particle size and other factors affecting edge tightness, therefore, have a significant effect on thickness swelling under submerged conditions. For instance, a combination of very thin flakes in a sized board of higher densities will create a sizable delaying effect on water penetration.

The environment in which particleboard is stored has a marked effect on the time required to attain maximum swell. The span of time needed is a matter of days when the board is submerged in 20°C. water. When high humidity (95 percent) is considered, it becomes a matter of months.

Under constant high humidity conditions, particleboard of usual thicknesses will continue to gain moisture for some two or three months before reaching equilibrium. Even longer periods are involved before thickness finally attains its fully expanded value (fig. 9–7). Some[6] show panel thickness playing a minor role in attaining equilibrium. Others,[7] using ASTM D1037–64, show that thicker boards (i.e., one inch) attain moisture equilibrium over a longer period (about two months) while thinner boards ($\frac{1}{2}$ inch or less) take about half as long. Cyclic exposure (dry, wet, dry) conditions appear to produce a greater thickness swelling compared to exposure at a constant comparably wet condition.

9–2. Linear Movements

Linear movements refer to hygroscopic dimensional change in the plane of the panels in particleboard. In randomly distributed particleboard, linear change with fluctuating moisture content is quite small compared with thickness change. In most cases, this change is limited to less than 1 percent in soak tests. Hygroscopic linear change in particleboard is similar to that of plywood (in which cross-lamination provides the product with a strong measure of stability). In randomly distributed particleboard, the linear change is strongly affected by the movement of

9–7. *Moisture absorption and thickness swelling for particleboard subjected to 95 percent relative humidity (at room temperature) over a period of almost five months.*

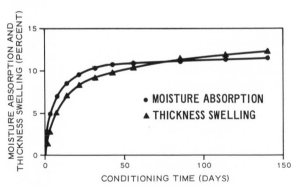

SOURCE: Liiri, O. 1961. Investigation of long-term humidification and the effect of alternate moisture variations on the properties of particleboards.
Redrawn in modified form by S.I.U. Cartographic Laboratory

9-8. *Linear dimensional expansion coefficient along (vertical) and across the grain (horizontal) for maple and walnut and particleboard.*

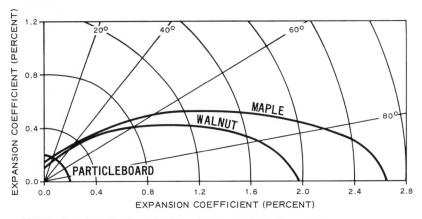

SOURCE: Suchsland, O. 1971. Linear expansion of veneered furniture panels. Redrawn in modified form by S.I.U. Cartographic Laboratory

wood along its grain. Studies have shown that linear change in flat-pressed particleboard ranges from 1.1 to 2.5 times the movements encountered in the parallel-to-grain direction of wood with particle geometry playing a very significant role.[8] Figure 9-8 illustrates linear stability by using a polar diagram comparing maple and walnut dimensional change along (vertical axis) and across (horizontal axis) the grain and in flat-pressed particleboard made of flakes. It is noted from this figure that particleboard is not only very stable linearly but that its movements are nondirectional in the plane of the panel. Along-the-grain stability imposes stabilization along the plane of the board. Nevertheless, linear movements are of interest: in some applications, long dimensions along the plane of the board are used for which the total dimensional change adds up to a large fraction of an inch. This becomes an important factor in structural and nonstructural uses. Should the panels be restrained from freely expanding and shrinking, self-imposed stresses will develop, which, under some conditions, cause buckling and the development of undesirable deformations. Both recoverable and unrecoverable movements are encountered in connection with linear movements although to a smaller extent compared to thickness (fig. 9-9). In most instances, board density appears to have insignificant influence on linear movement behavior.

In section 9-1 we noted that under exposure to high relative humidity conditions, thickness continued to increase for over five months. This, however, does not appear to be the case with linear movement. The

9–9. *Water absorption and linear movement for a 19-mm-thick, single-layer particleboard; (A) five cycles of exposure to high (95 percent) and low (30 percent) humidity and (B) five cycles of soaking (3 days) and drying at 30 percent relative humidity (11 days).*

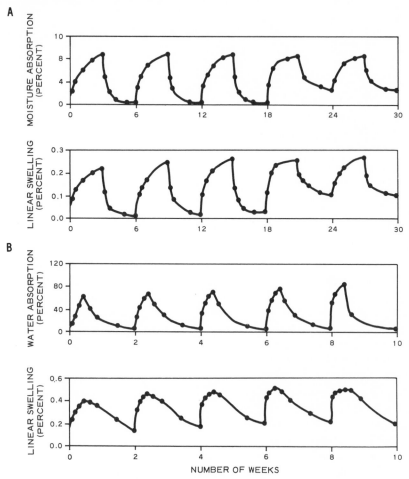

SOURCE: Liiri, O. 1961. Investigation of long-term humidification and the effect of alternate moisture variations on the properties of particleboards.
Redrawn in modified form by S.I.U. Cartographic Laboratory

9–10. *Average moisture absorption and linear swelling for particle-board with four different thicknesses and of single-layer construction when conditioned at 95 percent relative humidity over an extended period of time.*

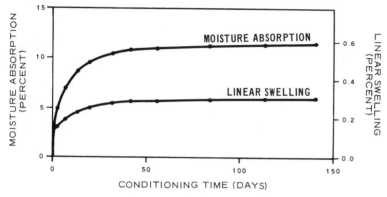

SOURCE: Liiri, O. 1961. Investigation of long-term humidification and the effect of alternate moisture variations on the properties of particleboards.
Redrawn in modified form by S.I.U. Cartographic Laboratory

latter, in a test conducted by Liiri,[9] reached its maximum value after approximately two months when exposed to constant 95 percent relative humidity at room temperature (fig. 9–10).

The long time intervals required to determine both thickness and linear movements of particleboard constitute a significant disadvantage. Such time requirements are likely to hamper research and development projects. Also, in routine quality control, it becomes extremely difficult, if not impossible, to specify dimensional changes in time with these techniques. As a result, various short-term tests have been proposed.[10] Some have used higher than room temperatures coupled with high (near 100 percent) humidities to reduce equalization time. Others have employed a vacuum-pressure-soak (VPS) method to achieve the same end result. In an abbreviated version of the VPS method, separate, matched specimens are taken, a set of which is oven-dried. The other set is simultaneously subjected to vacuum-pressure.[11] From the collected data, the dimensional changes are determined. It is worth noting that the linear measurements, taken shortly after oven-drying while the specimens are hot, include a certain amount of thermal expansion. However, such expansion appears to be very small compared to that induced by moisture content change and may, therefore, be neglected.[12]

9–3. Dimensional Stabilization

Much effort has been made to reduce dimensional instability—particularly that of thickness in particleboard. These attempts have been directed at the two basic sources of dimensional change in the panel:

(a) The hygroscopicity of the wood used.
(b) The release of built-in compressive forces brought about during manufacture.

The use of hydrophobing agents such as paraffin wax in particleboard is widespread and has been discussed earlier in this book (section 7–2). We have already noted that the use of wax is useful in short-term repelling of liquid water, with wax having practically no effect on water vapor flow.

Increasing the amount of resin content has also been employed to gain stability. Figure 9–11 illustrates the influence of this variable on soaking and high humidity conditions using urea-bonded particleboard of 40 pcf density made of Douglas-fir planer shavings. Wax sizing at the rate of 0.75 percent (solids, based on furnish weight) was used. When phenolic-bonded boards are utilized, the increase in resin content within practical ranges will also reduce thickness swelling as already shown (fig. 9–4). Resin catalyzation, too, appears to influence dimensional stability. When uncatalyzed resins are used, hardening of the binder will be affected. Kehr,[13] in a study using beech and pine, noted a considerable increase in thickness swelling when resins without hardeners were used. Hot pressing parameters such as temperature and time appear to have varying effects on dimensional stabilization. Press temperature, when changed through the practical range (300–450°F.), does not significantly affect dimensional change[14] while long press times can be deleterious to stability.[15]

Treatment of wood particles by various means has been attempted in order to reduce hygroscopicity. Treatments with heat,[16] polyethylene glycol,[17] impregnating phenolic resins, formaldehyde vapor, and acetylation[18] have been undertaken. Heating has been found to impart enhanced dimensional stability to solid wood.[19] Lehmann,[20] examining wood particleboard, subjected the particles to from fifteen to forty-five minutes of 400°F. heat, after which the particles were reconditioned to 5 percent moisture content at room temperature. Making particleboard from particles so treated, he found no worthwhile gains in thickness stabilization. On the contrary, with increasing treatment time, darkening occurred in particles—the likely factor responsible for definite reductions in board strength.

9–11. *The effect of two levels of resin content on thickness swelling of urea bonded, sized particleboard when subjected to water soak and high humidity.*

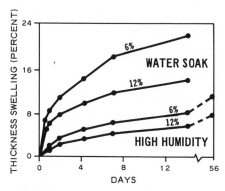

SOURCE: Lehmann, W. F. 1968. Durability of exterior particleboard.
Redrawn in modified form by S.I.U. Cartographic Laboratory

The stabilizing influence of polyethylene glycol as bulking agent in wood is by now well established. Undried particles have been sprayed with aqueous solution of this chemical prior to drying and resin blending. By using varying percentages of polyethylene glycol, considerable gain in thickness stability is achieved in resulting particleboard particularly when phenolic binders are employed. Using phenolic resins, the degree of interference with bonding appears to be very small or nonexistant as judged by strength losses. This is not, however, the case when urea binders are employed. The use of polyethylene glycol has its drawbacks: it requires extended press times and is subject to leaching which makes the beneficial effect of stability short lived.

Using impregnating phenolic resin indicates that a degree of dimensional stability is obtained with such resins.[21] Good stability has been obtained when a combination of impregnating phenolic resins and phenolic bonding resins are applied. The application of impregnating phenolic resin to moist particles in the amounts of 5 to 10 percent (in addition to 5 percent phenolic bonding resin during blending) has brought about a significant measure of stability to aspen particleboard. Boards so made reportedly provide a good degree of strength retention during weathering. Earlier work[22] also shows enhanced dimensional stability for high-density particleboard using impregnating phenolic resins.

The use of formaldehyde cross-linking has been shown to impart enhanced dimensional stability to wood. Dimensional change occurs in wood when water molecules enter or leave the cell wall thus increasing or reducing the distance between the swelling units. Stamm[23] and Tarkow[24] postulate that water molecules would be unable to enter between these swelling units if such units were held closely together by formaldehyde cross-links. Formaldehyde has been noted to form intermolecular acetals between the hydroxyl groups on carbon 1, 2, 4, and 6 of sugars with the others being too far apart to participate:

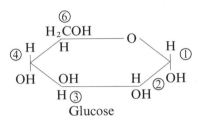

Glucose

In cellulose, it is possible that formaldehyde will link hydroxyls in adjacent chains—provided these chains are close together. For formaldehyde to be effective, it must be added in vapor form to an acid environment. Although good dimensional stabilization has been achieved with wood under certain conditions, its use in particleboard has not proved beneficial beyond any doubt.

Acetylation has also been used to reduce hygroscopicity in wood. This technique aims at eliminating the terminal hydroxyl groups in the cellulose molecule. A chemical reaction is called for such as the one between aliphatic acids (i.e., acetic acid) and the cellulose present in wood. In acetylation, acetyl groups partially replace hydroxyl groups. However, serious disadvantages of high cost, incompleteness of the reaction, loss of impact resistance, and detrimental effects on other useful properties have not encouraged the use of acetylation. Other means of replacing the hydroxyl groups exist, but none so far has shown any promise in dimensional stabilization of wood particleboard. Other approaches, e.g., modifications of the furnish pH and the extension of the adhesive with soluble blood have been tried to determine if any added stabilization can be attained. In particleboard, neither one of these techniques appears to have shown any promise. The use of tempering oil has also yielded no positive results.

A number of techniques have been attempted to relax compressive forces in particleboard thereby reducing thickness movements with change in moisture content. In one European practice, water is forced into the already pressed phenolic-bonded panels thereby attempting to relieve the built-in compressive forces. In this technique, the consolidated board is transferred to another press containing a water trough with foam rubber walls. When the pressure is applied (at 2 kg/cm²), the foam rubber becomes compressed forcing water rapidly (in about thirty seconds) into the board structure.[25] Moistened panels are stacked for a few hours, then dried in a kiln to a moisture content of 10 percent, and then subsequently sanded. This process attempts to eliminate any springback occurring after the first wetting—therefore reducing thickness change accordingly. In addition, it is claimed that the thickness of the board due to this treatment increases sufficiently to effect a 15 percent savings in material. The process can also incorporate wood preservatives into the panel along with the intended amount of water. Some manufacturers in the United States subject the pressed panels to high humidity followed by conditioning at low humidity to bring out the initial springback and thereby reduce subsequent thickness movements. Some underspray the panels and wet-stack them to achieve the same objective.

The use of heat in relaxing built-in compressive forces in phenolic-bonded board has been examined. For the heat treatment to be effective, either the heating periods must be long or the temperatures high. Such severe treatments usually bring about adverse effects (e.g., discoloration, strength reduction). Figure 9–12 indicates the influence of short-term

9–12. *Effect of heat treatment on thickness swelling and weight increase of particleboard when exposed to 90 percent relative humidity and redrying at room temperature.*

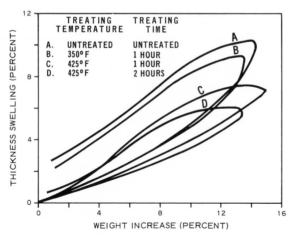

SOURCE: Suchsland, O., and Enlow, R. C. 1968. Heat treatment of exterior particleboard.
Redrawn in modified form by S.I.U. Cartographic Laboratory

9-13. *Effect of saturated steam treatment (360°F.) on thickness swelling of sized Douglas-fir particleboard.*

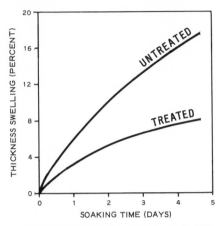

SOURCE: Heebink, B. G., and Hefty, F. V. 1968. Steam post treatments to reduce thickness swelling of particleboard.
Redrawn in modified form by S.I.U. Cartographic Laboratory

heating on a 44 pcf laboratory-made phenolic-bonded particleboard[26] indicating reductions in thickness movement when the boards were subjected to high humidity. Reductions were also obtained in twenty-four-hour water soak conditions. Although no appreciable losses in modulus of elasticity or internal bond strength were noted, severe drying caused serious board discoloration.

Subjecting phenolic-bonded panels to saturated steam has proven effective in controlling thickness swelling.[27] Significant reductions have been obtained when saturated steam at 360°F. is used (fig. 9–13). Moderate reductions in bending and internal bond strengths, however, are to be expected. This treatment does not appear to influence linear stability in any way.

In restraining linear movements, the use of overlays, particularly wood veneers, has been a common practice especially in the furniture industry. This, however, is a questionable proposition with flat-pressed particleboard, since very little improvement is achieved by overlaying. Figure 9–14 shows a polar diagram indicating the linear expansion coefficient for a symmetrical five-ply panel in which a particleboard core has been used. In obtaining this figure, the face consisted of .027-inch-thick walnut veneer with the crossband being made up of .029-inch-thick mahogany.

The core layer was a three-layer, 0.755-inch-thick particleboard. In figure 9–14 the vertical axis corresponds to the along-the-grain direction of face veneer and the horizontal axis indicates that of across the grain. The directional nature of the veneer sheet used is readily noted, as expected. However, neither the particleboard itself nor the overlaid panel shows directional behavior. The veneer overlay produces practically no added linear stability to the panel. Appreciable reductions in linear movements, on the other hand, are achieved in the machine direction when extruded particleboard is appropriately overlaid with the grain of the wood veneer parallel to machine direction in three-ply construction. Finally, a limited measure of dimensional stability may be achieved by various coatings (e.g., paint and synthetic sheets) as long as such coatings impede water or water vapor penetration into the particleboard.

9–4. Thermal Properties

Thermal conductivity of particleboard is of interest in both manufacturing and use. Solids, in general, transfer heat by conduction. In particleboard, the product is made up of wood interspaced by numerous air pockets. The wood itself, too, contains numerous small voids in its

9–14. *Linear expansion coefficient for a five-ply panel consisting of walnut facing, mahogany crossband, and three-layer particleboard compared to the expansion of its structural elements examined individually.*

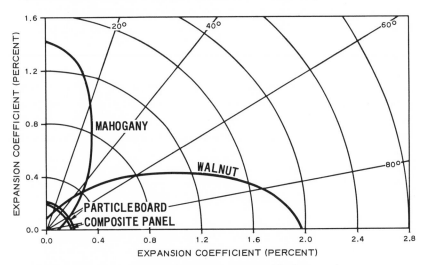

SOURCE: Suchsland, O. 1971. Linear expansion of veneered furniture panels. Redrawn in modified form by S.I.U. Cartographic Laboratory

structure. Radiation is the basic mode of heat transfer in such air pockets and voids. Convection effects are insignificant at normal temperatures. The presence of air pockets and voids makes heat transfer difficult thus creating inefficient heat conductors or, in effect, good thermal insulators. To form efficient heat insulators, a board with a large number of air spaces is preferred. In most cases, the size and number of air spaces in a given board is controlled by the overall density of the product (lower densities providing better insulators).

Thermal conductivity is measured in terms of units of heat transferred through unit thickness of the material during unit time per unit of area per unit of temperature difference between the two faces. Thermal conductivity is usually referred to by the symbol k which may be expressed in the British system of measurements or in the metric system. In practice, the former is often utilized; in research work, the latter. For instance, when the value of k is 0.70 for a given material in the British system, it means that for a one inch thickness, there is a heat transfer of 0.70 Btu[28] per hour per square foot for each Farenheit degree difference in temperature between its two surfaces. The direction in which heat is flowing makes a substantial difference in determining k values. Heat flow along the grain is 100 to 200 percent higher than across the grain.[29] In particleboard, too, direction of heat flow can be important. The value of k is different when heat is moving from one face to another compared to heat flow along or across the panel. Specific heat, c, is another parameter which refers to the quantity of heat required to raise the temperature of unit mass of the board by one degree. In the cgs units, c will be expressed in calories per gram per degree centigrade.

Ward and Skaar[30] measured k and c for different types of particleboard including an extrusion-process board. The apparatus used in the experiment was a modification of that proposed by Clarke.[31] It utilizes two twelve-by-twelve-inch specimens. The heat sink (six-by-six-inch) is composed of a highly conductive metal such as copper or aluminum and monitors the temperature. A three-inch-wide guard ring surrounds the heat sink. To determine the parameters k and c, the following equation was utilized:

$$k = \frac{dT_c}{dt} \cdot \frac{L(gcl + g'c'l')}{2(T_h - T_c)}. \qquad (9\text{–}7)$$

In this equation dT_c/dt is the rate of temperature rise of the heat sink (deg/sec), l the specimen thickness, g the density of the specimen (gm/cm^3), g' the density of the heat sink material (gm/cm^3), l' the thickness of the heat sink material, T_h the temperature of the hot plate (degrees C) and

T_c the temperature of the heat sink (degrees C). For equation 9–7 to be valid, the following condition must hold:

$$T_h - T_c = m$$

where m is a constant. The authors utilized two heat sinks with different thicknesses (l') in order to determine k and c from equation 9–7. Experimental data provide two sets of simultaneous equations yielding the value of the desired parameters. Both k and c were found to be functions of temperature as shown in figure 9–15 based on averages of five different flat-pressed particleboards with somewhat similar densities (about 0.7 gm/cm^3) and moisture contents. The type of adhesive did not appear to affect results. For extrusion specimens, the values of k and c would be different since the heat flow from one face to another is mostly along the grain of the particles. The parameters k and c are also a function of density since lower densities mean a greater number of air spaces in the board structure. Thus, thermal conductivity k would increase with increasing value of density. Figure 9–16 illustrates the influence of specific gravity on k for a lab-made Douglas-fir particleboard referred to as *standard* at the Forest Products Laboratory in Madison, Wisconsin. It is clear that less heat transfer (better insulation) will take place when lower temperatures and lower board specific gravities are involved. The layering in particleboard appears to play no significant role in thermal properties.[32] In practical applications, it is customary to use thermal conductivity values which are based on a mean temperature of 75°F. For general practical work, Lewis[33] recommends using a k value for each particleboard density category. For the low-density group (less than 37 pcf), recommended k value would be 0.94 and for high-density group (over 50 pcf), 1.18 Btu, per inch, per hour, per square foot, per degree F.[34]

9–5. Acoustical Properties

The control of noise has become increasingly important in the design of construction and dwellings. The source of such noise is either external to the building or internal. Therefore, the problem of noise control involves the isolation of external noise and the absorption of sound generated within. The physical laws applicable to isolation of sound are entirely different from those relevant to absorption. If a given material absorbs sound well, it will not necessarily isolate well. In fact, chances are that it will perform poorly in sound isolation if the given material is a good sound absorber. The reverse also holds true.

Sound absorption involves the conversion of sound energy into some

9–15. *Thermal conductivity and specific heat values as functions of temperature for five flat-pressed particleboards. Mean values (points) and regression lines are noted.*

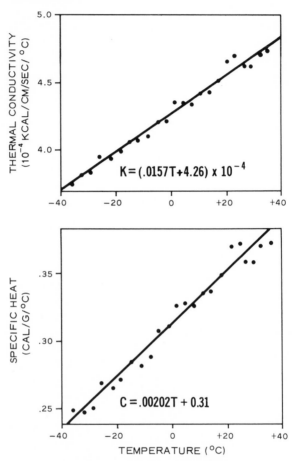

SOURCE: Ward, R. J., and Skaar, C. 1963. Specific heat and conductivity of particleboard as functions of temperature.
Redrawn in modified form by S.I.U. Cartographic Laboratory

other forms of energy such as heat as it passes through or strikes the surface of a material. On the other hand, sound isolation techniques rely on the mass and the stiffness of the material, although the principles involve complex relationships. A useful parameter in describing acoustical effectiveness of a material is referred to as the sound absorption coefficient.

This parameter is a measure of the absorbed fraction of sound energy striking the surface. Its value, however, is not constant depending on test conditions. The following variables will affect the coefficient value and therefore must be specified:

(a) Frequency of the sound.
(b) Angle of incidence of sound waves striking the surface.
(c) Method of specimen mounting.

9-16. *Effect of specific gravity on thermal conductivity for a urea bonded particleboard at four different temperatures.*

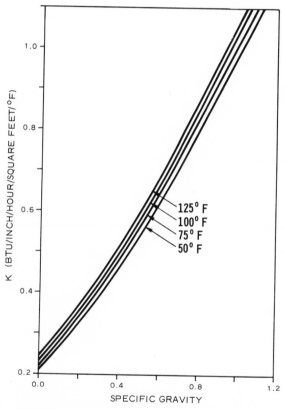

SOURCE: Lewis, W. C. 1967. Thermal conductivity of wood-base fiber and particle panel materials.
Redrawn in modified form by S.I.U. Cartographic Laboratory

9–17. *Frequency-absorption coefficient relationships for a ½-inch thick particleboard (made of flakes with 40 pcf density). Results for perforated hardboard have been shown for comparison.*

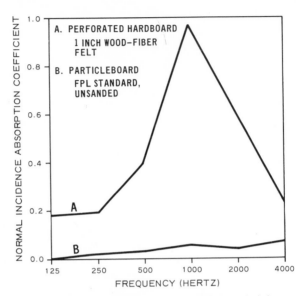

SOURCE: Godshall, W. D., and Davis, J. H. 1969. Acoustical absorption properties of woodbase panel materials.
Redrawn in modified form by S.I.U. Cartographic Laboratory

Figure 9–17 illustrates the influence of frequency on the absorption coefficient for normal incidence, with environmental humidity (30–80 percent) playing no significant role when tested at room temperature.[35] This figure compares the absorption coefficients for a *standard* Douglas-fir particleboard to those obtained for perforated hardboard. Much of the difference in this figure is due to the perforations in hardboard, since porous materials provide functional drag reducing sound energy into heat. Perforations in acoustical ceiling tiles make them far more efficient sound absorbers compared to identical tiles lacking such perforations. Materials with smooth surfaces will have low absorption coefficients regardless of material from which they are made. If particleboard could be made with rough or perforated surfaces, it would yield higher coefficient values compared to those presented in figure 9–17. This, for example, explains why excelsior board is being widely used for its desirable acoustical effects: because it consists of a loose fibrous mat with a relatively rough perforated surface.

Notes

1. Suchsland, O. 1966. Some performance characteristics of "interior" and "exterior" type particleboard.

2. Each cycle in figure 9–2 includes 3 weeks at 95 percent relative humidity followed by 3 weeks at 30 percent relative humidity.

3. Moduli of elasticity and rupture and shear strength are significantly affected. Internal bond, too, is reduced although to a lesser extent.

4. Suchsland, O. 1966. Some performance characteristics of "interior" and"exterior" type particleboard.

5. Keylwerth, R. 1958. On the mechanics of the multi-layer particleboard.

6. Liiri, O. 1969. The most recent developments in the wood particleboard line.

7. Heebink, B. G., and Hefty, F. V. 1968. Steam post treatments to reduce thickness swelling of particleboard.

8. Heebink, B. G., Kuenzi, E. W., and Maki, A. C. 1964. Linear movement of plywood and flakeboards as related to the longitudinal movement of wood.

Turner, H. D. 1954. Effect of size and shape on strength and dimensional stability of resin-bonded wood-particle panels.

9. Liiri, O. 1961. Investigation of long-term humidification and the effect of alternate moisture variations on the properties of particleboards.

10. Endicott, L. E., and Frost, T. R. 1967. Correlations of accelerated and long-term stability tests for wood-based composite products.

Heebink, B. G. 1967. A look at degradation in particleboards for exterior use.

Saums, W. A., and Turner, H. D. 1961. Accelerated test for measuring lateral dimensional change of wood base panels.

11. Heebink, B. G. 1967. A procedure for quickly evaluating dimensional stability of particleboard.

12. The coefficient of thermal expansion (the increase in length per unit of length for a temperature rise of 1°F.) in particleboard has been measured at 5×10^{-6} to 6×10^{-6}.

13. Kehr, E. 1964. Investigation of the suitability of various types and categories of wood for the manufacture of particleboard. Part 3.

14. Lehmann, W. F. 1964. Retarding dimensional changes in particleboards.

15. Kehr, E. 1964. Investigation of the suitability of various types and categories of wood for the manufacture of particleboard. Part 3.

16. Anon. 1959. Dimensional stability seminar.

Stamm, A. J., and Baechler, R. H. 1960. Decay resistance and dimensional stability of five modified woods.

17. Stamm, A. J. 1959. Effect of polyethylene glycol on the dimensional stability of woods.

18. Stamm, A. J., and Cohen, W. E. 1956. Swelling and dimensional control of paper. Part 2: Effect of cyanoethylation, acetylation and cross-linking with formaldehyde.

19. Lofgren, B. E. 1957. Continuous and discontinuous heat treatment and humidification of hardboards.

Ogland, N. J. 1957. The heat treatment of hardboard.

Seborg, R. M., Tarkow, H., and Stamm, A. J. 1953. Effect of heat upon the dimensional stabilization of wood.

Stamm, A. J. 1959. Dimensional stabilization of wood by thermal reactions and formaldehyde cross-linking.

20. Lehmann, W. F. 1964. Retarding dimensional changes in particleboards.

21. Haygreen, J. G., and Gertjejansen, R. O. 1971. Improving the properties of particleboard by treating the particles with phenolic impregnating resin.

22. Burrows, C. H. 1958. Floor tiles from planer shavings.

23. Stamm, A. J. 1959. Dimensional stabilization of wood by thermal reactions and formaldehyde cross-linking.

24. Tarkow, H., and Stamm, A. J. 1953. Effect of formaldehyde treatments upon the dimensional stabilization of wood.

25. Liiri, O. 1969. The most recent developments in the wood particleboard line.

26. Jack pine and aspen flakes were used respectively for face and core layers (three-layer construction). The face was blended with 12.5 percent and core with 7 percent resin using 0.5 percent wax.

27. Heebink, B. G., and Hefty, F. V. 1968. Steam post treatments to reduce thickness swelling of particleboard.

28. Btu is the abbreviation for the "British thermal unit." It refers to the amount of heat required to raise the temperature of one pound of water by 1°F.

29. Ward, R. J., and Skaar, C. 1963. Specific heat and conductivity of particleboard as functions of temperature.

30. Ibid.

31. Clarke, L. N. 1954. A method of measuring the thermal conductivity of poor conductors.

32. Lewis, W. C. 1967. Thermal conductivity of wood-base fiber and particle panel materials.

33. Ibid.

34. One Btu in. per hour, square foot, °F. equals 0.1240 kg cal per hour, meter, °C. or 1.442 milliwatts per cm, °C., etc.

35. Godshall, W. D., and Davis, J. H. 1969. Acoustical absorption properties of wood-base panel materials.

Moldings and Portland Cement-bonded Products

10-1. Definitions and Particularities of Moldings

Moldings may be essentially defined as wood products of various shapes made from a mixture of wood particles and synthetic binders. These products are molded into the desired shapes under the influence of heat and pressure by utilizing specially constructed dies (also known as molds or tools). In many processes, the wood particles used range in size from 8 to 140 mesh with the resin often in powder form although liquid binders are sometimes utilized. The resin content in moldings ranges from under 5 percent to some 35 percent or more, with the molding temperature ranging from about 270°F. to some 550°F. The applied pressure, too, ranges widely from some 200 psi to 4500 psi—depending on the process and the desired density.

When the resin content is relatively high, it becomes difficult to distinguish between plastics in which wood fillers are used and wood particle moldings. To allay this difficulty, some have defined moldings as wood-resin products in which the wood component can visually be readily identified or in which the resin content remains below 25 percent.[1] In the wood particle moldings classified as "plastics" by the definition given, specially processed wood flour or sander dust (smaller than 80 mesh) is used. Powdered resins employed for this purpose are urea, melamine, urea-melamine, or phenol-formaldehyde at percentages ranging from 25 to 40 by weight.

10–1. *A few examples of molding products.*

Photograph by Rip Stokes

From what has been noted, particles of up to 8 mesh may be utilized in moldings with resin content falling below 25 percent. This definition raises the question of how different moldings are compared to particleboard. The latter is also made from wood particles and a synthetic binder produced under pressure and heat. Although there are basic similarities between wood moldings and particleboard, the differences remain.

Wood particleboard is made in flat panel form possessing parallel surfaces. A molding often has unsimilar contours on the various sides. Thus, products designed to have other than flat surfaces, are classified as moldings. Repressed particleboard in which one surface is embossed, therefore, is not considered a particleboard but a molded product. Furthermore, the molded products are usually pressed to higher densities than particleboard, with some exceptions. In many cases, the resin content in moldings is often larger than that used for particleboard and the wood particle size employed is smaller in size. Molded products, unlike particleboard, are pressed into such shapes and contours that no further reshaping is usually required. A molding can be produced in such a way that grooves, notches, pilot holes, and overlays are already accounted for once the product emerges from the press. This will have obvious labor saving advantages by minimizing or altogether eliminating the successive processing steps.

Moldings may be engineered such that high material economics are achieved. For instance, the thickness of a given molded product can be varied so that higher stress or abrasion areas possess a higher density value compared to areas of low stress. Moldings are also produced in such a way that, similar to particleboard, they have a differential density profile over the thickness so that surfaces possess higher densities. Graduation over the product thickness with respect to particle size is similarly possible in moldings.

The variety of shapes that are molded is extremely large. Relatively flat, simple surfaces to highly intricate shapes are molded without difficulty —provided the correct choices of process and materials are selected. Figure 10–1 shows a number of examples.

10–2. Properties of Moldings

There are a number of properties of relevance which affect service-ability and appearance. The properties most often determined are flow (see section 10–3), modulus of rupture, and dimensional stability. In addition, such properties as cure cycle, water absorption, impact strength, and screw holding characteristics are considered important. A number of raw material and processing variables affect molding properties:

(a) Type of wood furnish.
(b) Particle geometry.
(c) Furnish moisture content.

(d) Type of resin.
(e) Resin content.
(f) Type and amount of additives.
(g) Blending techniques.
(h) Furnish bulk factor.
(i) Molding pressure.
(j) Molding temperature.
(k) Press closing time.
(l) Cure time.
(m) Materials and techniques employed in mold release.

Product density, as in particleboard, directly influences most properties—particularly strength. The density of moldings is generally a good indicator of its bending strength, hardness, abrasion resistance, and impact strength. Higher densities mean higher strength, hardness, and abrasion resistance characteristics. However, there are disadvantages associated with higher densities: high-density products will be higher in weight, which is generally considered undesirable in most applications. In addition, high-density moldings have a greater tendency to warp. The density distribution within a given molded part is important. For instance, in cases where the surface is composed of a dense, resin-rich layer compared to a similar product of uniform but equivalent density, the former is likely to have a lower impact strength.

It is difficult to specify accurately the interrelationships of various physical properties for all varieties of molded products. However, it is possible to provide indications of the influence of density on various properties of parts of simple and uniform shapes. Brockschmidt,[2] in attempting to find these relationships, molded a series of test discs. Ponderosa pine particles (100-mesh) with a moisture content of 5 percent were blended with powdered phenolic resin (270-mesh plus). The amount of resin added was 12 percent based on the total resin-wood mixture. The blended mix was hot-pressed into 4-inch diameter discs with a thickness of $\frac{1}{2}$ inch (controlled by stops). The die temperature used was 310°F. and the molding pressure 2,600 psi. The specific gravity of the molded discs was controlled by the amount of material placed into the die.

Increased specific gravity was found to retard moisture take-up. Figure 10–2 illustrates that the amount of water take-up was sharply and linearly reduced as the specific gravity of the disc approached the value of 1.0.[3] Increasing the specific gravity of the sample beyond this value brought further reductions in water absorption at a reduced rate. Figure 10–2 further shows that the screw holding power and impact strength are increased as the specific gravity of the product is increased.[4]

10-2. *(A) Specific gravity and water absorption relations for two types of resin formulations. (B) Interaction of particle mesh size and molding specific gravity on the screw withdrawal resistance of the molding. (C) Impact strength as a function of specific gravity for four different furnishes:*
(a) 20-mesh particles containing 16 percent resin,
(b) 40-mesh particles containing 20 percent resin,
(c) 100-mesh particles with 12 percent resin, and
(d) 100-mesh particles with 20 percent resin.
(D) Press cure times versus molding composite densities.

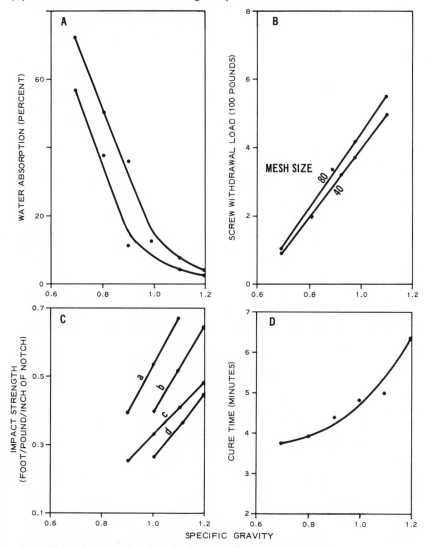

SOURCE: Brockschmidt, K. H. 1960. Fundamentals of molding wood particles.
Redrawn in modified form by S.I.U. Cartographic Laboratory

Figure 10–2 shows that particle mesh size slightly affected the results of this property.

Cure time is an important parameter in much the same way as it is in particleboard. Longer cure times would mean longer press times and lower productive capacity. The definition of cure time varies somewhat, depending on various authors and processes involved. Generally, it is taken as the time required to produce a well-cured molding without inducing any blisters, blows, or bulges. This technological parameter is affected by a number of factors including the specific gravity of the molding and the die temperature employed. In wood moldings, denser materials require longer cure times (fig. 10–2). This is due to the fact that there is more material to be heated and cured in denser products. Thicker moldings, too, would have a similar influence on molding time due to the existence

10–3. *Temperature-cure time relationships for moldings.*

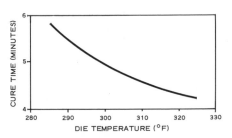

SOURCE: Brockschmidt, K. H. 1960. Fundamentals of molding wood particles.
Redrawn in modified form by S.I.U. Cartographic Laboratory

of more material to be heated and a greater distance of heat travel involved. Elevating the die temperature reduces, up to a limit, the cure time required for any given condition (fig. 10–3). This situation, too, is similar to that existing in the hot pressing of particleboard. Resin content of the furnish appears to be independent of cure time (fig. 10–4).

Dimensional stability of moldings is of prime importance and is influenced by a number of factors. I have already noted that increasing specific gravity reduces water take-up in a short soak time. However, dimensional stability depends on a number of other factors, e.g., the natural properties of the wood used, particle orientation (of longitudinal grain direction), the degree of wood compression, and the level of resin content. Particle geometry appears to have an influence: Gatchell and Heebink,[5] comparing ripsaw sawdust and hammermilled flakes, noted that the latter particles produce moldings of poorer thickness stability but better linear stability. Examining the influence of resin content on dimensional stability of moldings (containing 6–18 percent resin) made from Douglas-fir and yellow birch particles, it was noted that increased levels of resin content translated into a more stable product. Beyond 12 percent, further stabilization effects were noted although at a reduced rate. The authors believe that this phenomenon is due to the coverage of the wood particles by the resin at around 12 percent. Beyond 12 percent, it is believed that further addition of resins would not be of value in improving the dimensional stability of the product.

The influence of furnish moisture content upon mold closing time and product density is similar to that noted for particleboard. Namely, hot pressing the furnish of higher moisture content will result in faster closing

10–4. *Amount of resin-cure time relationships for moldings.*

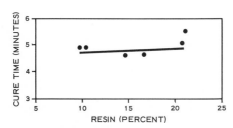

SOURCE: Brockschmidt, K. H. 1960. Fundamentals of molding wood particles.
Redrawn in modified form by S.I.U. Cartographic Laboratory

times and a differential density profile with the surfaces becoming more densified. In wood particle molding, furnish moisture content will also affect its flow characteristics. Molding at higher moisture levels (over 8 percent) can create problems. Large quantities of entrapped volatiles will tend to lengthen resin curing time and produce fractures and blisters upon pressure release. High levels of furnish moisture will also tend to promote excessive resin flow and cause problems in mold release.[6] It is reemphasized that the release of volatiles in molding is often very slow and therefore the overall moisture content of the blended furnish must be kept at a low level. This is a major reason for using binders in the powder form in most molding operations. Using powder resins also has the additional advantage of keeping the blending and molding equipment relatively clean.

The type of resin selected has a strong bearing on the technological and product properties of the moldings. When urea-formaldehyde is used, products with light and stable colors are generated, although the moisture resistance and durability will be low. When phenolic resins are utilized, improvements in the properties just noted will be obtained but at an extra cost due to the higher price of the phenolics. Phenolic resins also produce a product of darker color. This may cause coloring problems when dyes are to be incorporated into moldings. Phenolic resins produce good flow and other technological properties which result in adequate strength, screw holding, and surface hardness.

Melamine-formaldehyde produces moldings of equal integrity compared to those made using phenol-formaldehyde. Melamine-formaldehyde has an additional advantage: it does not bring about problems associated with color darkening. Thus, when moisture resistance is desired in moldings into which dyes and pigments are to be incorporated, melamine resins may be used. Furthermore, good surface gloss and hardness are achieved with the use of melamine binders. However, melamine resins have a lower flow capacity than phenolics,[7] a disadvantage when producing products of intricate shapes.

10-3. Flow

Flow is an important property of the furnish and refers to its ability to move under the influences of heat and/or pressure at right angles to the applied force. High flow is essential when molding products of intricate shape. Conversely, only shapes with simplified configurations can be molded should the furnish exhibit low flow characteristics. There are a number of parameters that affect flow. The most prominent single factor is the amount of resin content in the furnish. Low resin furnishes will exhibit poor flow and vice versa. Press pressure, too, affects flow: higher pressures mean increased flow.

In a study designed to determine the extent of influence that the two previously mentioned parameters (resin content, pressure) exert on the flow characteristics of wood-resin mixtures, Brockschmidt[8] utilized a preform 51 mm in diameter and placed it in a 102 mm diameter die. Pressure was applied at room temperature upon which the material began to flow. Once the furnish movement was completed, the diameter of the formed disc was measured to determine the amount of flow. Table 10-1 shows that both increased resin content and pressure promoted flow, with the contribution of the latter being of limited degree. This and other research show that mixtures with less than 20 percent resin content exhibit poor flow characteristics.[9] On the other hand, the cost of resin in moldings

10–1. *Inter-relationship of resin content and mold pressure on flow*

Resin Content (percent)	Mold Pressure (psi)	Distance of Flow (mm)
10	650	56.5
10	1300	64
10	1950	70
10	2600	77
20	650	72.5
20	1300	83.5
20	1950	90
20	2600	95
35	650	90
35	1300	97.5
35	1950	100
35	2600	102 (max.)
50	650	102 (max.)
50	1300	102 (max.)
50	1950	102 (max.)
50	2600	102 (max.)

SOURCE: Brockschmidt, K. H. 1960. Fundamentals of molding wood particles.

will be a major factor in determining whether or not the manufacture of a given line of products will be advisable.

To measure flow, various techniques have been used. In his research, Brockschmidt determined the diameter of the disc after the completion of pressure application as an indicator of flow (starting with a constant diameter preform). This constitutes a simple means, but it does not give any indication of density uniformity within a given disc. Gatchell and Heebink,[10] dealing with low resin blends, compared center and outside weights of plugs taken from the molded disc made from a cone-shaped mat (fig. 10–5). The flow index, f, was defined as follows:

$$f = 100 \frac{W_1}{W_2}$$

in which W_1 is the weight of an outside plug and W_2 the weight of the center plug. The flow index values would then be low under poor flow conditions with f approaching 100 percent as greater flow is achieved.

Yukna and coworker,[11] dealing with moldings made from sawdust and urea-formaldehyde, measured flow by dividing the molded article into a number of segments (fig. 10–6). The weights of outer and inner segments were then examined. In making a molding, a given amount of

10–5. *Molding die used in the study by C. J. Gatchell and B. G. Heebink.*

MALE PLATEN

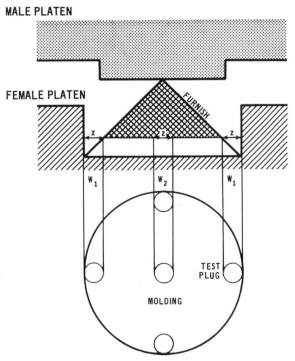

SOURCE: Gatchell, C. J., and Heebink, B. G. 1964. Effect of particle geometry on properties of molded wood-resin blends.
Redrawn in modified form by S.I.U. Cartographic Laboratory

furnish was poured into the die in such a way that an even height was achieved in the mold. This was done to obtain a practically identical amount of furnish over the die. If no flow was to take place, no differential weight distribution would exist, i.e., the weight would be about the same over all segments of the product. Conversely, should flow take place during the molding operation, some migration of material would take effect from one part of the molding to another resulting in uneven weight distribution. Taking these facts into consideration, one can write the theoretical density for the external and internal segments of the molding assuming the situation in which no flow takes place. Mathematically, the external theoretical density can be written as:

$$\gamma'_e = \frac{p \; w_e}{w \; v_e}. \tag{10–1}$$

In this equation, γ'_e is the theoretical density for the external segments, p is the total weight of the molding, w is the total width of the segments, w_e is the weight of the external segments and v_e is the volume of the external segments. The parameters in equation 10–1 are expressed as follows (fig. 10–6):

$$p = p_e + p_i \tag{10–2}$$
$$w = w_e + w_i \tag{10–3}$$

in which

$$p_e = p_1 + p_4 + p_5 + p_8.$$
$$p_i = p_2 + p_3 + p_6 + p_7$$
$$w_e = w_1 + w_4 + w_5 + w_8$$
$$w_i = w_2 + w_3 + w_6 + w_7.$$

In equation 10–1, v_e can be expressed as

$$v_e = v_1 + v_4 + v_5 + v_8$$

In the same manner, the internal theoretical density γ'_i is written (fig. 10–6):

$$\gamma'_i = \frac{p\, w_i}{w\, v_i}. \tag{10–4}$$

The parameters p and w are the same as those mentioned in equations 10–2 and 10–3. The term w_i, too, has just been defined. The term v_i, on the other hand, is defined as:

$$v_i = v_2 + v_3 + v_6 + v_7.$$

10–6. *Segmentation of a molding in order to study flow characteristics.*

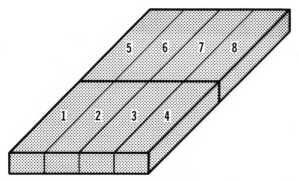

SOURCE: Yukna, A. D., Bankiyeris, Ya., and Ziyedin'sh, I. O. 1965. The flow of particle-glue mixture in pressing profiled parts from pulverized wood.
Redrawn in modified form by S.I.U. Cartographic Laboratory

To determine flow, equations 10–1 and 10–4 are compared with actual density (after flow) which can be defined as:

$$\gamma_e = \frac{p_e}{v_e} \qquad (10\text{–}5)$$

$$\gamma_i = \frac{p_i}{v_i} \qquad (10\text{–}6)$$

in which the parameters p_e, p_i, v_e, and v_i are defined the same way as before. Thus, flow in the external segments, f_e, is:

$$f_e = \gamma_e - \gamma'_e \quad (\text{gm/cm}^3) \qquad (10\text{–}7)$$

and in internal samples, f_i, will be

$$f_i = \gamma_i - \gamma'_i. \quad (\text{gm/cm}^3) \qquad (10\text{–}8)$$

Expressing f_e and f_i in percentages, we get:

$$f'_e = \frac{100(\gamma_e - \gamma'_e)}{\gamma'_e} \quad (\%) \qquad (10\text{–}9)$$

$$f'_i = \frac{100(\gamma_i - \gamma'_i)}{\gamma'_i}. \quad (\%) \qquad (10\text{–}10)$$

Using the same raw materials and manufacturing procedures, Yukna and coauthor find that the amount and pattern of flow depend on the shape of the article to be molded. Table 10–2 illustrates a number of profiles pressed at a range of densities and the value of flow, in percent, obtained for the external and internal segments of the specimen. In this table, the plus sign denotes gain in material while the minus sign denotes loss.

10–4. Portland Cement

An inorganic binder which has found commercial use in the field of certain wood particle panels is portland cement.[12] This binder is used in the manufacture of products intended for service in structural applications which may or may not be load bearing. Portland cement contains four basic chemical elements—namely, silicon, aluminum, iron, and calcium, usually quoted as oxides. Other large numbers of elements with possible presence in cement are unessential and are represented by relatively small quantities. Of the basic iron and aluminum elements, the existence of only one of the two is required in the production of

10-2. *Flow properties of a number of molding configurations*

Profile	Density gm/cm³	Number of Specimens per Profile	f'_e (percent)	f'_i (percent)
	0.56	8	−1.03	+1.64
	0.67		−0.98	+1.52
	0.78		−3.67	+4.28
	0.88		−2.44	+2.35
	0.59	6	−4.86	+9.99
	0.66		−3.91	+7.75
	1.08		−4.93	+9.23
	0.62	6	−9.76	+19.70
	0.98		−10.62	+21.47
	0.65	8	−23.81	+24.62
	1.01		−29.55	+30.00
	0.60	6	−9.09	+17.00
	0.73		−8.28	+22.15
	1.15		−14.80	+18.25
	0.57	6	−6.13	+12.43
	0.89		−5.16	+8.93
	0.99	6	−0.11	+0.82
	0.59	6	−9.56	+21.50
	0.94		−9.83	+18.54

SOURCE: Yukna, A. D., Bankiyeris, Ya., and Ziyedin'sh, I. O. 1965. The flow of particle-glue mixture in pressing profiled parts from pulverized wood.

cement. Table 10-3 lists the quantitative composition of some of the raw materials used to manufacture cement. Of those listed, limestone is the most widely utilized raw material source.

The manufacture of portland cement involves the incorporation of raw materials to form a homogeneous mixture, the burning of the mixture in a kiln, and the grinding of the product (along with a small proportion of gypsum) into a fine powder. There are two processes of cement manufacture based on whether or not the raw materials are ground and mixed in a dry or wet state. In recent times, the dry process appears to be more preferred—due, in part, to its lower energy requirements for burning even though the wet process is usually credited with a more accurate

10–3. *Raw materials and their composition used in cement manufacture*

Material	SiO_2 (percent)	Al_2O_3 (percent)	Fe_2O_3 (percent)	CaO (percent)	MgO (percent)
Limestone	0.78	0.28	0.26	53.69	0.66
Cement rock	13.77	5.00	1.61	41.49	2.06
Marl	0.15	0.27	0.19	54.70	0.88
Oyster shell	1.50	——	——	53.80	0.25
Clay	54.24	29.12	5.72	6.75	——
Tufa	53.76	22.88	6.84	5.41	1.88
Blast furnace slag	37.37	11.81	0.65	45.50	2.75
Iron ore	9.02	4.76	72.30	4.76	1.17
Bauxite	2.00	58.83	21.99	0.22	0.03
Diaspore	21.94	55.87	9.13	——	——
Sand	61.35	18.46	4.82	5.22	2.72

mixing. In the dry process, the received raw materials are first crushed followed by drying in rotary dryers. At this stage, they are proportioned and ground to reduce the material particle size and directed to the ground raw material storage silos. The rotary kiln in which the cement is burned, is a long, inclined cylinder to which the material is fed at the upper end. The material to be burned travels slowly to the lower end of the kiln as the latter continuously rotates on its axis. The energy source, such as pulverized coal, is blown into the kiln where it is ignited. The material fed into the kilns undergoes a series of reactions within the kiln environment forming a granular product called clinker. Upon exit, the clinker enters a cooler. The clinker is then routed to an intermediary storage or sent directly to the grinding mills. A small proportion of gypsum is added during grinding to keep the cement product from setting. Storage silos hold the finished product which is, in time, dispensed, packed, and marketed. The wet process of cement manufacture differs from the dry process in that water is used to form a slurry out of the raw material mix which is eventually sent to the kiln to be burned and converted into clinker. In this process, wash mills are utilized to break up and incorporate the raw materials if the latter consists of chalks and marls. These materials are fed in the required proportions along with sufficient quantities of water to form a slurry of creamy consistency. However, when limestones and shales (which are considerably harder) are being used, reduction cannot take place in wash mills. In this case, the raw material and water are directed to crushers, which are required to generate the desired slurry. The latter is subsequently fed into special storage tanks where means are available to keep the slurry homogeneous.

Both standard and high early strength portland cement are utilized in bonding wood particle panels. The production of the latter grade of cement—an early 1900s development—involves a number of modifications. Some of these modifications include finer grinding, different mixing of the raw materials and adjustment of the makeup of materials, usually resulting in a higher lime content compared to the standard grade. High early strength cement may be so ground as to yield a particle surface area ranging from 2400 to 3000 cm^2/gm. Standard portland cement (type 1) has a particle surface area of 1600 cm^2/gm. Generally, there is a spectrum of properties having no definite dividing lines from standard to high early strength portland cements.

Ready-to-use portland cement is composed of a number of inorganic compounds which set and harden in the presence of water. Table 10–4 shows the result of the analysis of a standard, type 1 portland cement indicating that tricalcium silicate ($3CaO.SiO_2$) is the most dominant single component. All components in portland cement are anhydrous and when in contact with water, form hydrated compounds. Hydration of tricalcium silicate starts promptly as soon as it comes in contact with water with the reaction producing calcium hydroxide. The process of hydration, as it continues, results in liberation of calcium hydroxide crystals. The hydration of dicalcium silicate ($2CaO.SiO_2$) advances only slowly with some lime liberated during the reaction. The hydration of tricalcium aluminate ($3CaO.Al_2O_3$) is rapid, resulting in the production of hexagonal plate and needlelike crystals which continue to increase in size and amount with the advance of the hydration process. The relatively high heat of evolving hydration is capable of rendering the material almost dry. Tetracalcium aluminoferrite ($4CaO.Al_2O_3.Fe_2O_3$) hydrates upon contact with water, although the speed of reaction is not as fast as is the case for tricalcium aluminate. When tetracalcium aluminoferrite is mixed with water to form a paste, it shows good crystal formation in a day. Thus, it is believed that the bond with portland cement is generated through a crystallization process with a mass of interlaced crystals involved. The bond made up of hardened cement has an ordered crystalline structure. It is believed

10–4. *Analysis of a standard, type 1, portland cement*

Compound	Amount (percent)
Tricalcium silicate ($3CaO.SiO_2$)	48
Dicalcium silicate ($2CaO.SiO_2$)	27
Tricalcium aluminate ($3CaO.Al_2O_3$)	12
Tetracalcium aluminoferrite ($4CaO.Al_2O_3.Fe_2O_3$)	8

that the hardening of cement is due to the intergrowth of interlocking crystals, although other widely different theories on the subject have been advanced.[13] The interlocking crystals theory pictures the hardened cement as consisting of folded and buckled foils and ribbons. The crystals grow outward from the cement particles making contact with each other. At some of these contacts, the touching crystals will have similarly oriented lattices, enabling them to intergrow by a solid state reaction, thus becoming welded together. The number and strength of these welds increase as hydration proceeds, thereby increasing the strength of the bond and, of course, of the panel. The growth of these crystals fills the voids between the original cement grains and the wood particles. Interspersed in this mass, there still remain voids and capillaries filled by water that is lost upon drying.

10-5. Wood-Cement Bonding

Bonding wood particles with portland cement has not been without problems. Many wood species possess a number of organic constituents which have deleterious effects on the wood-cement bond. In fact, in some instances, the influence of such substances is so severe that the cement binder is inhibited from setting at all regardless of the amount of time allowed. This inhibitory effect appears to be more pronounced in hardwoods compared to softwoods due to the existence of the type of constituents that retard or prohibit cement setting.[14] Early research has placed the responsibility on hardwood hemicellulose.[15] In addition, be it a hardwood or softwood, other substances such as starches, tannins, sugars, and some phenols have been blamed for having an inhibitory effect on wood-cement bond formation. Glucose has been particularly singled out for adversely influencing cement setting.[16] Cellobiose in decayed softwoods has also been shown to be strongly inhibitory. Weatherwax and Tarkow,[17] in studying wood-cement systems, defined an inhibitory index as follows:

$$I = \frac{100(t\text{-}t')}{t'}$$

in which I signifies the index in percent, t' is the time in hours required for uninhibited cement to reach its maximum temperature with t the time needed for inhibited cement. The temperature referred to here is generated by the heat of hydration which attains a maximum value before it begins to drop. Figure 10-7 shows two examples, one for an uninhibited and the other for an inhibited cement. Generally, the sooner the maximum temperature is reached, the quicker the cement will set.

10–7. *Typical temperature rise and the heat of hydration for uninhibited (curves A and B) and inhibited (with 4.2 mμ of glucose per gram of cement, curves C and D) portland cement.*

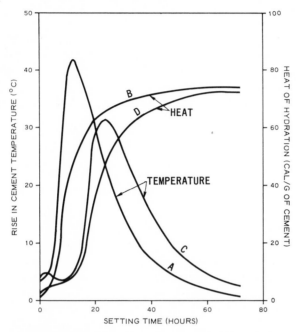

SOURCE: Weatherwax, R. C., and Tarkow, H. 1964. Effect of wood on setting of portland cement.
Redrawn in modified form by S.I.U. Cartographic Laboratory

Weatherwax and Tarkow, examining a number of metabolic products of fungal decay, found that cellobiose and glucose, using the inhibitory index defined above, were the most inhibitory (fig. 10–8). The inhibitor concentration, *m*, for cellobiose and glucose was related to *I*, respectively as follows:

$$I = 24.8(m-0.49) + 0.458(m-0.49)^4$$
$$I = 24.8(m-0.49).$$

In these equations, the inhibitor concentration is in micromole (equivalent of glucose) per gram of cement.

Other compounds such as carboxymethyl cellulose and cellulose could also cause setting problems. Tannins, especially of the hydrolyzable type, and a number of organic acids of higher molecular weight may act likewise. Phenolic groups such as those present in lignin or bark, in

10–8. *Setting time with four types of inhibitors: (A) cellobiose, (B) glucose, (C) gluconic acid, and (D) citric acid.*

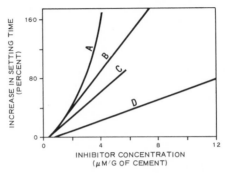

SOURCE: Weatherwax, R. C., and Tarkow, H. 1967. Effect of wood on the setting of portland cement: Decayed wood as an inhibitor.
Redrawn in modified form by S.I.U. Cartographic Laboratory

moderate amounts, are not believed to affect portland cement setting.[18] Terpenes, too, do not appear to adversely influence hardening.[19] Some have argued that the time of year when the wood is cut has relevance on setting: spring-cut wood, they contend, appears to retard setting due to the presence of higher quantities of starches and other inhibitory substances.[20] However, research on southern pines has indicated no substantive differences in setting between the spring-cut and winter-cut wood. Heartwood as opposed to sapwood is inhibitory[21] in southern pines and probably other species as well.

Davis,[22] in an experiment consisting of six experimental runs, notes that stain caused by the fungus *Ceratocystis pilifera* (Fr.) C. Moreau promoted setting as measured by the heat of hydration. Davis postulates that the reason lies in utilization of soluble constituents by the stain fungus thereby removing them from acting on the cement hardening and bond formation. This is in contrast to the action of wood destroying fungi which produces soluble byproducts of inhibitory nature.

To reduce the inhibitory effects of wood on cement setting, various measures have been taken. One of these consists of extracting sugars and other water soluble materials by hot water.[23] The wood particles are soaked in hot water for a period of time. Soak duration depends on a number of factors such as the temperature of hot water and the amount of inhibitory extractives present. The addition of various chemical additives, too, have been found useful in reducing setting time. A major compound used in this connection is calcium chloride. Clare[24] states that calcium chloride solutions of 1–3 percent are necessary to neutralize the effect of 0.1 percent sugar in the soil-cement mixtures. Further increases in the level of calcium chloride, however, have been found to reduce strength in such mixtures.[25] Among other compounds, aluminum sulfate in aqueous solution has also been used to forestall the inhibitory effect of sugar when added at 4–5 percent rate.[26] Caustic soda at concentrations of 0.1–1 percent has been utilized on a number of difficult to bond oriental species[27] with the authors reporting good results.[28] The caustic soda in this case was believed to remove inhibitory substances in sufficient quantities to provide adequate strength to the wood-cement bond.

It is probable that, in many instances, the wood raw material source can contain a portion which may be inhibitory on cement setting. This may occur, for instance, when some decayed wood is included. Weatherwax and Tarkow,[29] in dealing with this problem, propose an equation which determines the inhibitory index of the wood mixture from the weight fraction of the decayed wood and the inhibitory indexes of the sound as well as the decayed wood:

$$\log I_m = (l-w_d)\log I_s + w_d \log I_d.$$

In this equation, I_m is the inhibitory index of the wood mixture in percent, w_d the weight fraction of decayed wood, I_d and I_s are the inhibitory indexes of decayed and sound wood respectively. Figure 10–9 illustrates how I_m is increased as the value of w_d assumes higher values. In examining

10–9. *The influence of the weight fraction of decayed wood* W_d *on the inhibitory index,* I_m, *using exclusively the wood of southern pines.*

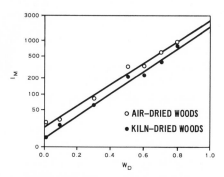

SOURCE: Weatherwax, R. C., and Tarkow, H. 1967. Effect of wood on the setting of portland cement: Decayed wood as an inhibitor.
Redrawn in modified form by S.I.U. Cartographic Laboratory

this figure, Weatherwax and Tarkow suggest that an inhibitory index of 40 should be considered the maximum value beyond which objectionable inhibitory effects may be encountered.[30]

10-6. Portland Cement-bonded Products

A popular wood particle cement product consists of excelsior board used for a number of structural applications—particularly roof decking. For this product, long and stringy excelsior particles (some ten inches or more in length) must first be generated. The wood species utilized are usually low-density woods, e.g., aspen and poplar. Various types of southern pines—particularly loblolly, slash, and longleaf, are also used. Low-density hardwoods form the raw material in the Great Lakes and northeastern states while the southern pines are primarily used in the Deep South.

The excelsior machine, whether vertical or horizontal, utilizes short length bolts of less than two feet in length with many using nineteen-inch-long pieces. The wood raw material obtained may be in the form of roundwood of convenient length (in many cases eight feet) which is first debarked and then slashed into the short lengths needed. In some cases, tree-length roundwood is hauled to the mill, debarked, and slashed before reduction to excelsior particles. The diameter of the wood material used ranges from four to eleven inches with the larger material split before reduction into particles.

The moisture content of the wood raw material has a direct influence on the quality of reduction in the excelsior machine. When the material is too dry, an unacceptably large proportion of fines will be generated. If it is too wet, the shearing action of the excelsior machine knives will make a "lazy cut," which reduces production. The optimum moisture content varies, depending on species: for southern pines this is generally in the 32–35 percent range. For northern hardwoods such as aspen, a moisture content in the 20–22 percent range is recommended. Once the tree is cut and slashed, it is air-dried for a period of time depending on location. For southern pines, this period may range from eight to ten weeks. In northern climates using light-density hardwoods, it may range from six to twelve months.

Figure 10–10 outlines the process of manufacture for excelsior board. In this figure, it is noted that the cement and particles are mixed with the assistance of some water. Once the particles enter the mixer, the cement and water are added, while paddles contained in the mixer continue to agitate the material. Calcium chloride is also added at this stage in the manufacturing process. Once the wood particles, cement, and $CaCl_2$

10–10. *Basic steps of excelsior board manufacture.*

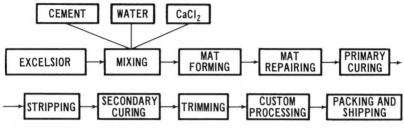

Drawing by S.I.U. Cartographic Laboratory

are blended, the material is ready to be formed into mats. Most excelsior boards made for decking have a density of about 36 pcf. For this density, experience has established a mat-to-board thickness ratio of about 3:1. This indicates that, for example, if a final board thickness of two inches is desired, at the forming stage, a six-inch mat is constructed. Mat forming generally consists of the blended material falling loosely over a caul plate which consists of an exterior-grade plywood sheet. Once the needed thickness is built up, the mat moves out of the forming machine. Formers used in excelsior board industry often operate less accurately compared to those used in a particleboard plant. It is often necessary to repair a badly distributed mat, a task usually done manually using visual judgment.

The mats, as they are formed, are directed to the press and are, in turn, pressed to the final thickness through the use of stops which also act as edge supports. In reality, these stops often consist of four lumber strips placed on four sides of the mat between the two plywood caul plates. In a given press load, a number of panels ranging from approximately thirteen to seventeen or more are consolidated depending on the thickness of the product being pressed. A thickness ranging from $1\frac{1}{2}$ to 3 inches is commonly produced.

It takes a number of hours for the cement bond to develop the necessary strength. Obviously, the press cannot be tied up with a load for such long periods. It is therefore, a common practice to apply the necessary restraint through portable, lightweight equipment which in some cases may consist of steel straps. Often a more substantial means made up of steel bars and tightening devices forming clamps are used. Once the restraint is tightened, the press is opened, and the stack is removed and carried to a primary curing room. The stack may spend from six to eight hours in the primary curing room during which the particle-cement bond continues to gain strength. After the necessary hours have passed, the restraint is removed. The boards can now be handled without damage. The panels are then stickered for good airing. The operation of removing and stickering the panels is referred to as stripping, which usually requires two men. The stickered stacks are then carried and placed in the outdoor condition, usually in the mill yard, where secondary curing continues for at least 7–8 days before any further processing takes place. The panels are trimmed, custom-processed, packed and shipped. The custom-processing referred to here may include custom-cutting, surface painting, and custom-patching. Carbide-tipped saws are used in cutting.

Due to the porous nature of excelsior board, good acoustical (sound absorption) and thermal properties are obtained. The cement binder used also tends to retard combustibility, thus enhancing fire resistance. However, the product is not designed to carry significant loads. Figure 10–11 shows a typical roof decking application.

Portland cement has also been utilized to manufacture particle panels of higher density compared to that of excelsior board. The particles

10–11. *An example of roof decking application for excelsior board.*

Courtesy Concrete Products Co., Brunswick, Ga.
Photograph by Schwarm and Sheldon, Inc., Fort Lauderdale, Fla.

utilized are not long and stringy. In these products, the particles are often flakes and are deposited either in random, single layer fashion or may be layered and oriented (in face layers) in order to provide the panel with high bending strength in the direction of surface particles.[31] In any event, the particles are mixed with cement and water, after which they are formed into mats, using specialized forming machines, some of which orient the surface particles in the machine direction. The mats are deposited on metal caul plates and subsequently directed to a cold press which will consolidate them to the desired density (usually greater than one). The need for long-term clamping also exists here. After some curing, the clamp is released and the panels dried. The panels are then trimmed and processed (cut to desired size, painted, etc.). Both standard and high-early-strength portland cement grades may be utilized with the latter grade providing panels which possess a moduli of rupture and elasticity of some 4000 and well over a million psi respectively at a panel density of little over one tested at air-dried conditions.

The latter type of wood particle-cement product, under development for many years in the United States,[32] is being currently manufactured in a number of foreign countries—including Japan and Switzerland. In the manufacture of these products in Japan, wood flake measuring $1\frac{1}{2}$–2 inches long, $\frac{1}{32}$–$\frac{3}{16}$ inch wide, and about 0.010 inch thick are cut and mixed in continuous mixers with cement and water. Cement and wood are mixed by weight at the rate of two parts cement to one part wood. Once the mats (three-layer, with faces oriented) are formed, a number are stacked and end-fed into a cold press consolidating them to a stop. Restraining clamps are then applied and held some 16 hours before the stack is opened. To accelerate drying, the panels are dried in a mechanical dryer by using lightly heated air.[33] The boards produced are 3 feet wide and either 6 or 9 feet long and $\frac{1}{2}$ inch thick. The panels are mostly factory painted and are usually intended for exterior application (sheathing) in the field of light construction (fig. 10–12). The product is being used in both vertical and horizontal situations. The interest generated in this type of wood particle composite not only lies in the interesting choice of raw materials used, it is also brought about as a result of other advantages, i.e., fire and termite resistance as well as weatherability. However, higher densities compared to resin-bonded particleboard are involved. The panels, in the usable density range (0.9–1.4 gm/cm^3) can be sawed and nailed. Above the density of 1.4 gm/cm^3, the sawing and nailing become difficult.[34] The panels possessing densities below 0.9 gm/cm^3 will have inferior exposure resistance and low strength.

In order to produce a product of high quality in the correct density

10–12. *One-inch-thick wood particle-cement panels used as siding on a house in Yokohama, Japan.*

Courtesy Elmendorf Research, Inc., Palo Alto, California

range, careful attention must be paid to the ratio of wood to cement and wood to water used. Improper ratios not only influence panel properties adversely, they can also make processing difficult. In the mixing operation, the necessary water is first added to the particles and is followed by cement addition. This water is adequate to start cement hydration and commence bond formation. It also assists in coating the particles uniformly with cement.

Notes

1. Pagel, H. F. 1967. Molding wood particles.
2. Brockschmidt, K. H. 1960. Fundamentals of molding wood particles.
3. The relationship shown in figure 10–2 was obtained when the specimen discs were immersed in 77°F. temperature water for a period of 24 hours.
4. Screw holding was determined by using a No. 6–$\frac{3}{4}$ inch round head wood screw inserted into a drilled $\frac{7}{64}$ inch hole pulled by a testing machine. ASTM standards for evaluation the impact resistance of phenolic molding compounds were used. In this test, the specimen is subjected to a blow or impact. Record is made of the energy required to break the specimen in foot pounds per inch of specimen notch cut out previously.
5. Gatchell, C. J., and Heebink, B. G. 1964. Effect of particle geometry on properties of molded wood-resin blends.

6. Graho, J., Jr., and Williams, F. H. 1954. The manufacture of molded articles from resin-bonded granulated wood.

7. Anon. n.d. Granulated wood molding.

8. Brockschmidt, K. H. 1960. Fundamentals of molding wood particles.

9. Gatchell, C. J., and Heebink, B. G. 1964. Effect of particle geometry on properties of molded wood-resin blends.

Patterson, T. J., and Snodgrass, J. D. 1959. Effect of formation variables on properties of wood particle moldings.

10. Gatchell, C. J., and Heebink, B. G. 1964. Effect of particle geometry on properties of molded wood-resin blends.

11. Yukna, A. D., Bankiyeris, Ya., and Ziyedin'sh, I. O. 1965. The flow of particle-glue mixture in pressing profiled parts from pulverized wood.

12. An English stonemason named Joseph Aspdin was granted a patent in 1824 for a new type of cementitious material which he called "portland cement." The color of the product after hydration reminded him of the limestone of the Isle of Portland.

13. Lea, G. M. 1956. The chemistry of cement and concrete.

14. Weatherwax, R. C., and Tarkow, H. 1964. Effect of wood on setting of portland cement.

15. Christensen, L. E., and Lyneis, R. G. 1948. The effect of sugar and wood extracts on the properties of portland cement mixtures.

16. Clare, K. E. 1956. Further studies on the effect of organic matter on the setting of soil cement mixtures.

17. Weatherwax, R. C., and Tarkow, H. 1964. Effect of wood on setting of portland cement.

18. Kleinlogel, A. 1950. Influences of concrete.

Weatherwax, R. C., and Tarkow, H. 1964. Effect of wood on setting of portland cement.

19. Kleinlogel, A. 1950. Influences of concrete.

20. Biblis, E. J., and Lo, C. F. 1968. Effect on setting of southern pine cement mixtures.

21. Clare, K. E. 1956. Further studies on the effect of organic matter on the setting of soil cement mixtures.

Weatherwax, R. C., and Tarkow, H. 1964. Effect of wood on setting of portland cement.

22. Davis, T. C. 1966. Effect of blue stain on setting of excelsior cement mixtures.

23. Biblis, E. J., and Lo, C. F. 1968. Effect on setting of southern pine cement mixtures.

24. Clare, K. E. 1956. Further studies on the effect of organic matter on the setting of soil cement mixtures.

25. Kleinlogel, A. 1950. Influences of concrete.

26. Schmidt, L. 1958. Selection of mineralizer and the method of mineralization in making cement fibrolite.

27. *Betula platyphylla sukatchev* var. *japonia* HARA; *Larix leptolepis* GORDON; *Plerocarya rhoifolia* SIEBOLD & ZUCCARINI.

28. Yashiro, M., Kawamura, Y., Sasaki, K., and Mamada, S. 1968. Studies on manufacturing condition of wood wool-cement board. Part 1: Trial manufacturing of wood wool-cement board with unsuitable species.

29. Weatherwax, R. C., and Tarkow, H. 1967. Effect of wood on the setting of portland cement: Decayed wood as an inhibitor.

30. Badly decayed southern pine wood yields an inhibitory index value in the 2000–11000 range while kiln-dried sapwood may give a value as low as 5.

31. Patent (U.S.) 1965. Patent No. 3,202,743.

Patent (U.S.) 1966. Patent No. 3,271,492.

Patent (U.S.) 1969. Patent No. 3,478,861.

32. This product was developed and refined at Elmendorf Research Laboratories, Inc. of Palo Alto, California. Patent (U.S.) 1954. Patent No. 2,697,677.

33. Wood, W. A. 1968. Cement-coated particles used in Japanese board.

34. Patent (U.S.) 1966. Patent No. 3,271,492.

Glossary

Unit Conversions

Bibliography

Index

Glossary

This glossary appears in identical form in both Volumes 1 and 2.

Abrading Plate. Grinding plates attached to the revolving disks of attrition-type fiberizing units.

Abrasion Resistance. The surface property which implies resistance to being worn due to friction or rubbing action.

Abrasive Debarker. Equipment debarking roundwood by frictional processes.

Absorption. The act of one substance being assimilated by another.

Activator. See Catalyst.

Addition Polymer. A polymer which does not split out water as it advances from monomeric to polymeric state; example, polyesters.

Additives. Compounds added to the furnish in order to impart certain properties to the product.

Adhesion. The property which causes one substance to hold on to another.

Adhesive. Substance which causes materials to stick to each other through surface attachment.

Adsorption. A type of adhesion which develops at the surface of solids or liquids while in contact with another medium.

Aging. Allowing the given material to be exposed to certain conditions undisturbed for usually an extensive period of time.

Alkyd. A condensation resin produced by a polybasic acid and a polyhydric alcohol.

Annual Ring. The increment of wood generated in a single growth year and seen in the cross section of a piece of wood; same as growth ring.

Attrition Mill. Reduction unit utilizing revolving disks with abrading plates to fiberize lignocellulosic raw materials. It is also sometimes used to blend binders with the stock while the latter is being reduced.

Bagasse. Lignocellulosic leftover of sugar cane after the juice has been taken out.

Bamboo. The lignocellulosic cane of a number of treelike tropical grass species possessing a hollow, jointed structure.

Bark. Outer tissue of the trunk and branches (outside cambium) of trees; it is composed of inner living and outer dead bark.

Bark Extract. An aqueous extract of various bark species possessing a significant quantity of complex, tanninlike constituents which are reactive to formaldehyde and have potential value in the manufacture of resins.

Barking. Process of removing the bark from roundwood; same as debarking.

Bast Fibers. Fibers belonging to the secondary phloem.

Batch. A more or less specific, finite quantity of a material such as furnish used to manufacture particleboard.

Batch Process. Process utilizing batches as opposed to one using continuous streams of material.

Beater Mill. Secondary reduction unit reducing the size of already produced particles.

Bending Strength. Strength capability of particleboard as it is subjected to bending test.

Binder. Common term for natural or synthetic adhesives.

Bleed-Through. Discoloration caused by penetration of adhesive through the face veneer once the latter is glued to the particleboard substrate.

Blender. Blending apparatus.

Blending. Incorporation of adhesives and additives with the particle furnish.

Blister. A type of failure characterized by a breakdown in adhesion. Existence of high pockets of moisture is a common cause. The bubblelike failure may occur in veneer overlaying or painting of substrates or surfaces having excessive moisture.

Blow. Partial separation of the board beneath its surface.

Blow Agents. Low-blowing substances such as trichlorofluoromethane which are used to assist foaming in the production of particleboard with polyurethane foams.

Bond. The uniting force holding the particles together in particleboard.

Bonding. The action to effect a bond.

Bonding Efficiency. The amount of strength obtained in particleboard per unit of resin usage.

Bound Water. Water in wood held by the cell wall.

Bowing. Curved distortion of panel material describing flatwise longitudinal warping.

Brittleness. Tendency of coatings to crack or flake. Also applies to the tendency of the board to break off readily upon impact.

Buffers. Compounds added to synthetic resins to extend their working life. Such compounds (example: tricalcium phosphate) slow down the rate of polymerization at room temperature.

Bulk Density. Unit weight of furnish in loose, uncompacted state. Very thin or fibrous furnish normally will have low bulk density.

Bunker. A storage for particle material along the production line.

Captive Plant. A manufacturing setup producing materials for another operation located in the same locality.

Catalyst. A substance intended to accelerate hardening in adhesives; same as hardener, accelerator, or activator.

Caul Plate. A flat metal plate on which blended furnish is deposited and pressed. In the production of excelsior board, caul plates may consist of exterior-grade plywood sheets.

Cellulose. A complex carbohydrate occurring in large quantities in higher plants and possessing the empirical formula $(C_6H_{10}O_5)_n$.

Cell Wall. A multilayered wall in the mature wood cell surrounding lumen.

Cement. An inorganic binder containing four basic chemical elements: silicon, aluminum, iron, and calcium and generally meaning, in this book, portland cement.

Cereal Straw. Lignocellulosic stalk of grain species.

Chip. A chunky lignocellulosic particle of various sizes.

Chipboard. A particleboard panel made of chiplike particles. This term is sometimes used instead of *particleboard*. The latter is the preferred term.

Chipping. Reduction of solid wood to particles. Also refers to the failure of paint film or other coatings with the material separating in flakes or chips. Chipping further refers to separation of pieces from the board itself.

Closing Time. The time period from the moment the loose mat is placed in the hot press to the moment when full pressure or stops are reached.

Cohesion. The force holding together the molecules of the same substance.

Compatibility. The ability of two or more substances to mix or contact each other without separation or reaction. No detrimental effects would result in mixing compatible materials.

Compression Ratio. Refers approximately to the ratio of the densities of the wood species to the particleboard from which it is made.

Compression Strength. Particleboard strength exhibited by forces tending to shorten the test specimen.

Conditioning. Bringing into equilibrium the moisture content and temperature of particleboard with those of a desired environment.

Coniferous Wood. Wood belonging to coniferous tree species; same as softwoods.

Contact Area. The proportion of total particle area in the board structure which is in firm contact.

Control Chart. A statistical chart monitoring various product properties showing whether or not they are within the prescribed boundaries.

Core. The inner layer of particleboard.

Core Stock. Particleboard used for overlaying. The overlay may consist of wood veneer, plastics, etc.

Cupping. Warping in the panel such that it assumes a concave shape. The opposing panel edges remain approximately parallel to each other.

Cured Panel. Refers to a solidified particle panel in which the binder has largely hardened.

Cut-to-Order. Sawing the standard panel sizes into various shapes and dimensions requested by secondary manufacturers; same as cut-to-size.

Daylight. Refers to the opening in platen-type presses. For instance a twenty-daylight press signifies one in which there are twenty openings.

Debarking. See Barking.

Deckle Frame. A framelike structure used to support the furnish edges during forming and pressing.

Delamination. Partial or total separation along the plane of the board.

Demixing.   Segregation of the various segments of a dry mix.

Density.   The weight of the material per unit volume expressed in either grams per cubic centimeter, kilograms per cubic meter or pounds per cubic foot.

Density Profile.   Density differentiation over the thickness of the board; usually highest at the surfaces and lowest in the innermost layer.

Die.   The shaped metal structure used to consolidate moldings. It usually has a male and a female part although it is not always easy to distinguish the two.

Dilutability.   The ability to become diluted with water.

Dimensional Stability.   Refers to the lack of dimensional changes in particleboard as the environmental humidity fluctuates or as the product is wetted with water.

Doctor Roll.   A segment of the roll coating equipment insuring uniformity in the coating thickness being applied.

Dryer.   Equipment utilizing heat to bring the particle moisture content to acceptably low levels.

Drying.   The process of evaporating excess moisture from the particle furnish.

Drum Sander.   A sanding machine possessing rolls or drums covered with sandpaper.

Durability.   The ability of particleboard to retain its original, desirable properties over an extended period of time.

Dust.   Very fine granules of wood material existing in the furnish or generated by the sanding operation.

Edge Banding.   Gluing lumber strips to the edges of particleboard before overlaying.

Elephant Grass.   A cattail (*Typha elephantina*) growing over a wide region of the world.

Emulsifying Agents.   Compounds used to hold wax particles in an emulsified state.

Emulsion.   A short term used to refer to wax emulsions.

Enamel.   A general term applied to free-flowing pigmented varnishes, treated oils, or lacquers. Enamels usually dry to a glossy or semiglossy hard finish.

Excelsior.   Long stringy particles used in the manufacture of excelsior board.

Excelsior Board.   A cement-bonded, relatively low density panel made up of excelsior particles and used mostly for roof decking although there are a number of other structural applications.

Extender.   Materials used to increase coverage of synthetic resins and improve certain properties. Cereal flour and blood are often used as extenders.

Exterior Particleboard.   A particle panel product, usually phenolic resin-bonded, intended for exposed or semiexposed outdoor applications.

Extractives.   A large variety of compounds found in wood which are not essential parts of the wood substance.

Extruded Board.   Particleboard made by the extrusion process and characterized by particles being largely at right angles to the board surfaces.

Extruder.   The long, parallel hot plates (extrusion die) used to consolidate and harden the board in the extrusion process.

Extrusion Process. A process of particleboard manufacture in which the loose, blended furnish is forced into two hot parallel plates known as the extrusion die.

Face Veneer. Veneer used as an overlay for the exposed surface.

Fiber. A relatively long, thin element of wood or other lignocellulosic materials.

Fiber Bundle. A large number of fibers which have not completely separated into individual elements.

Filler. A usually inert product applied to the particleboard surface in order to fill the pores before the desired finish is applied.

Fines. Very small, dustlike particles.

Fire Retardant. Chemical compound(s) incorporated into the board structure in order to impart some degree of fire resistance to the product.

Fissure. A separation (crack) occurring across the face in mats formed by caulless processes.

Flake. A particle characterized by relatively long length and very small thickness.

Flakeboard. A resin-bonded particleboard made up of flakes.

Flaker. A reduction unit utilizing cutter knives and producing flakes.

Flame Spread. A term used to describe the ease with which flames spread over the particleboard surface.

Flapreg. A high-density particleboard made of resin-impregnated particles.

Flash. The extra material squeezed out once the male and female parts of a molding die come together.

Flash Dryers. Particle dryers designed to dry the incoming furnish within a few seconds.

Flatness Ratio. The ratio of particle length to particle width.

Flat-Press Board. A particleboard manufactured by the flat-press process.

Flat-Press Process. Utilizes platen-type presses to apply pressure perpendicular to board faces. Particles generally fall flat along the plane of the board.

Flax. An annual plant of the genus *Linum* with slender, erect stem and linear leaves.

Flaxboard. Resin-bonded particleboard made of flax shives.

Flax Shive. After retting and scutching (to remove linen fiber), the remaining stalks are reduced into particles called flax shives.

Flow. The ability of the furnish to move under pressure and heat in moldings.

Fluted Board. A thick, extruded-type particleboard exhibiting holes along the length and within the thickness designed to lower unit weight. The holes are generated by pipes centrally located between the two hot plates in the extrusion die.

Formaldehyde. HCHO

Formaldehyde Release. Dissipation of formaldehyde odor during manufacture and use of synthetic resin-bonded particleboard.

Forming. The action of depositing the blended, loose furnish into a mat of desired characteristics; same as felting, spreading.

Forming Machine. Equipment designed to do the task of forming; same as former, felter, spreader.

Free Moisture.   Water found in the wood cell cavities either in liquid or vapor form.

Fungicide.   Chemical compound(s) incorporated into particleboard designed to forestall fungal attack.

Furnish.   Loose particle stock.

Glue.   Same as adhesive or binder.

Graduated Board.   A particleboard made up of a graduated mat.

Graduated Mat.   A symmetrical mat in which the smallest particles are located on the surfaces with the coarsest at the innermost layer.

Granule.   A term used to describe a small, chunky particle.

Hammermill.   A reduction unit utilizing breaking, tearing, or grinding action.

Hardboard.   A generally thin fiberboard utilizing the binding action of natural lignin in standard grades. The density of hardboard normally exceeds 30 pcf. It is difficult to distinguish fibrous medium-density particleboard from hardboard.

Hardening Agent.   See Catalyst.

Hardwood.   Wood of broad-leaved trees.

Heartwood.   The dead inner area of a stem or a log often but not always readily distinguishable due to its darker color.

Heat of Hydration.   Heat generated by portland cement as it comes into contact with water and begins to harden.

High Density.   The class of particleboard with density exceeding 50 pcf.

High-Frequency Heating.   A technique in which electrical energy is utilized to generate heat by molecular friction.

Hog.   A reduction unit using impact-type impellers to reduce wood and other lignocellulosic materials into crude particles.

Hot Press.   A major piece of equipment in the plant which consolidates and hardens resin-bonded particleboard by subjecting the mat to pressure and heat. The press platens are heated by either hot water, hot oil, steam, or electricity. In some designs, it may also subject the thick mats to high-frequency heating.

Hot Pressing.   The act of pressing the mat in a hot press.

Hot Stacking.   Solid stacking of hot panels as they exit the hot press.

Hydraulic Debarker.   A debarker utilizing water under high pressure to remove the bark.

Hydrolysis.   Process of chemical decomposition either in the wood material or the binder.

Hygroscopic.   Characterizes ready absorption and retaining of moisture as the environmental humidity is increased. The reverse also holds.

Impregnation.   Introduction of chemicals, e.g., fungicides, into the manufactured board or particles. Also introduction of resin into the particles prior to consolidation in flapreg production.

Impurities.   Foreign matter, e.g., inorganic dust, metal objects, etc., mixed in with wood particles.

Imcompatible.   This term is used to indicate that a given material cannot be mixed with another without deleterious effects.

Inhibitory. Describes a material or substance which will interfere with setting of portland cement.

Inorganic Binder. A binder of inorganic origin; examples, portland cement and gypsum.

Insecticide. Chemical compound(s) incorporated into particleboard to forestall insect attack.

Intermediate Density. In particleboard this ranges from 37 pcf to 50 pcf; same as medium density.

Internal Bond. A common term used to describe the tensile strength of particleboard in the direction perpendicular to the surfaces of the board; same as tensile strength perpendicular to the surface.

Lacquer. A variety of film-forming materials that usually dry by evaporation with the film formed from the nonvolatile constituents.

Lacquer Thinner. An organic combination of solvents.

Laminate. A single component of a laminar structure of more or less distinct characteristics.

Lignin. A highly complex principal constituent of woody cell walls.

Lignocellulosic. Materials of plant origin whose principal constituents consist of lignin and cellulose.

Linear Change. Dimensional variations occurring along the plane of particleboard primarily due to hygroscopic effects; same as linear movement.

Loader. A device used to transfer mats into the hot press; same as press loader.

Logging Residue. Wood residue left after the logging operation has been completed. This includes branches, cull logs, stumps, etc.

Low Density. A particleboard whose density is less than 37 pcf.

Mat. A mattress of particles deposited on caul plates, steel belts, etc., in a predetermined manner by the forming machine.

Mat Preheating. Raising the temperature of the mat somewhat above the room temperature with the principal objective of shortening hot-pressing time.

Medium Density. See Intermediate Density.

Melamine. It is a white, crystalline chemical (2, 4, 6-triamino-1, 3, 5-triazine) used to manufacture plastics and resins.

Melamine Resins. Synthetic adhesives (belonging to the amino resin class) which are condensation products of melamine and formaldehyde. Same as melamine-formaldehyde. They have high heat resistance and stable color.

Methanol. Methyl alcohol (CH_3OH).

Modulus of Elasticity. A strength parameter calculated from test data.

Modulus of Rupture. Maximum bending strength in pounds per square inch.

Moisture Content. The amount of water present in the wood material or particleboard expressed in terms of the percent of the oven-dry weight.

Molding. Resin-bonded product of usually various shapes and profiles; same as mouldings.

Multi-opening Press. A press possessing more than one daylight; same as multi-daylight press.

Neat Resin. A resin intended for use on "as is" basis.

Oil Contamination. Oil droplets striking the particle surfaces during drying.

Opening. See Daylight.

Outer Layer. Refers to the outermost layer of particleboard; same as surface layer.

Oven-Dry Weight. The weight obtained after the material has been drying long enough in an oven (105°C.) so that no further weight loss is noted.

Overlay. A sheet material of various origins (wood, plastics, metals, etc.) glued to particleboard substrates.

Papyrus. A fibrous tall sedge (*Cyperus papyrus*) with a triangular stem.

Particle. A general term describing a fragment of wood or other lignocellulosic materials ranging in size from a single fiber to a coarse chip such as that used in the pulp industry.

Particleboard. A resin-bonded product made basically of wood or other lignocellulosic particles and produced in panel form. Particle geometry may vary from fibers to large chips. Particleboard may consist of a single or a number of layers.

Particle Geometry. Refers to particle shape and size.

Particle Reduction. Reducing particles in size.

Particle Separation. Categorizing particles by size using screens or air classifiers; same as particle classification.

Peel-Off. Partial separation occurring in the plane of the mat during manufacture in caulless processes.

Pentachlorophenol. A chemical preservative also commonly referred to as Penta.

Phenol. A chemical also known as hydroxybenzene or carbolic acid with its compact formulation being C_6H_5OH. In pure form it is a crystalline substance with specific gravity of 1.072.

Phenolic Resins. Synthetic adhesives produced as a condensation product of phenol and an aldehyde such as formaldehyde.

Phloem. Inner bark.

pH Value. A number denoting the degree of acidity or alkalinity.

Polymer. Chemical compound(s) formed by polymerization and basically consisting of repeating structural units.

Polymerization. A chemical reaction where two or more molecules combine to form a larger molecule of repeating structural makeup. This term is also used to describe curing or hardening of synthetic resins.

Polystyrene Resins. Synthetic resins characterized by white to light yellow color and produced by polymerization of styrene. The properties of these resins can vary widely depending on the conditions under which polymerization takes place.

Polyurethane Foam. The result of reaction between an isocyanate and a compound containing active hydrogen atoms such as polyester or polyether.

Polyvinyl Acetate Resins. These polymeric thermoplastic resins are colorless, tasteless, and odorless.

Pot Life. The length of time an adhesive remains usable after the catalyst is added; same as working life.

Precure. The curing of a synthetic resin in blended material or mat before pressure application.

Prepress. A press designed to accomplish preliminary consolidation of the mat in order to reduce its thickness and increase its self-rigidity before entering the hot press for final consolidation and hardening.

Press Breathing. Partial or total reduction of press pressure during mat consolidation in the hot press before the pressure is reincreased; same as breathing.

Press Loader. See Loader.

Press Stops. Metal gauge bars inserted between press platens intended for board thickness control; same as stops.

Press Unloader. See Unloader.

Pressure Diagram. Pressure-time relationships during pressing.

Pulpwood. Roundwood of relatively small diameter usually utilized in pulp manufacture.

Radio-Frequency Heating. See High-Frequency Heating.

Recycling. Reutilizing material that has already been once manufactured into a product, put in service, and discarded.

Relative Humidity. Water vapor present in a unit air volume expressed in percent of water vapor required to saturate the same volume with temperature and pressure remaining constant.

Residual Moisture. Moisture remaining in particleboard upon its exit from the the hot press.

Residue. Wood by-products of various industrial activities such as slabs, trims, and veneer clippings.

Resin. A general term used to refer to synthetic adhesives.

Resin Blending. See Blending.

Resin Content. The amount of resin solids usually expressed as the percent of oven-dry weight of the particles.

Resin Hogging. This term is used to describe a situation when the particles take up a greater than intended amount of resin during the blending operation.

Roller Coating. A technique of applying coatings by using hard rubber or steel rolls.

Roundwood. Wood raw material in the round form.

Sander. Device equipped with abrasive drum, belt, or pad designed to impart smoothness to the surface or to bring board thickness to within close tolerances; same as sanding machine.

Sanding Allowance. Amount of extra thickness allowed for sanding.

Sandwich Construction. A panel construction with high-strength, dense faces, and low-strength, low-density core.

Sapwood. Outer section of a stem or log usually of lighter color compared to heartwood.

Sawdust. Small particles produced when wood is sawn. Also, this term may refer to small wood particles in general.

Screen. A device used to separate particles by size.

Screening. Action of particle separation using screens.

Screening Analysis. Scrutiny of particle size distribution through the utilization of screens.

Semifiber. A term occasionally used to describe partially defibrated furnish.

Shaving.   Relatively small particles produced as the by-product of planing operation; same as planer shavings.

Shelling Ratio.   A parameter related to the amount of particles used for surface layers compared to the entire board.

Shive.   See Flax Shive.

Show-Through.   A characteristic in which the particle substrate pattern shows through the overlay; same as telegraphing.

Size.   Refers to paraffin wax additive used to increase water resistance.

Slab.   A type of wood residue produced during primary log conversion.

Slenderness Ratio.   Ratio of particle length to particle thickness.

Sliver.   A relatively long, chunky particle.

Softwood.   Wood of coniferous trees.

Solids Content.   The percent of solid matter present in the adhesives or additives in terms of the total weight.

Specific Strength.   The ratio of a given strength parameter to the density.

Splinter.   Somewhat similar to sliver although less uniform in its dimensions.

Spreader.   See Forming Machine.

Springback.   Permanent dimensional increase either upon pressure release or water (liquid or vapor) absorption.

Stability.   Refers to lack of hygroscopic dimensional changes.

Steam Shock Effect.   This describes rapid steam transfer during hot pressing from consolidated and heated mat faces toward the core followed by quick temperature increase in the core.

Steel Belt.   A belt which acts as conveyer and caul plate in caulless processes.

Steel Cloth.   A metal cloth which acts as conveyer and caul plate in caulless processes. The meshed design assists quick cooling and reuse.

Storage Life.   Time period during which adhesives may be stored without degradation.

Strand.   A relatively large flake.

Stress.   Force applied per unit area.

Styrene.   A clear, colorless liquid also known as vinyl benzene (C_8H_8). It is used to manufacture polystyrene resins.

Sulfite Waste Liquor.   A complex by-product of the sulfite pulping process.

Surface Smoothness.   Refers to the smoothness of panel surface; opposite of surface roughness. The term *surface quality* often refers to surface smoothness.

Surge Bin.   In-production particle storage.

Suspension Dryer.   A type of particle dryer which suspends particles in a heated tube or cylinder in the course of drying.

Synthetic Resins.   Complex organic materials of semisolid or solid nature manufactured by reacting comparatively simple compounds. In the particleboard industry, this term is employed to refer to binders used.

Tack.   Slight stickiness with which blended particles can adhere to each other.

Tailoring.   Custom-formulation of resins meeting the requirements of a given situation.

Telegraphing.   See Show-Through.

Tensile Strength. Resistance of particleboard to tension.

Thermal Insulation. The tendency of particleboard to retard the passage of heat.

Thermoplastic. Applies to resins which soften upon heating.

Thermosetting. Applies to resins which harden upon heating.

Thinnings. Trees of usually small size cut to make adequate room (thus to permit better growth) for the standing trees.

Thixotropy. The property of some dispersions to become liquid upon shaking but coagulating again once left alone.

Three-Layer Board. Refers to particleboard possessing three distinct layers (of different particle sizes, resin contents, etc.) two faces and one core.

Unloader. Device used to remove the consolidated panels from the hot press; same as press unloader.

Urea. A prominent member of the nitrogen containing amino resins (NH_2CONH_2).

Urea Resin. Synthetic resin obtained by reacting urea with formaldehyde in the presence of acid or alkaline catalysts. It is, by far, the most common resin utilized in particleboard manufacture.

Vascular Cambium. A growing layer between wood and bark in trees; same as cambium.

Vehicle. The liquid segment of any coating or finish. Vehicle may contain various compounds in dissolved state.

Veneer. A thin sheet of wood.

Vinyl Acetate. A liquid with no color which is used to make vinyl resins. Its chemical formula is $CH_2CHCO_2CH_3$.

Vinyl Resin. Thermoplastic resin resulting from the polymerization of such compounds as polyvinyl acetate, polyvinyl chloride, and polyvinyl acetals.

Viscometer. Device specifying viscosity of liquids.

Viscosity. A liquid or gas parameter basically referring to its resistance to flow.

Warp. Any type of distortion in particleboard panels.

Water Absorption. The amount of water absorbed during soaking and expressed in percent of dry weight.

Water Repellent. Additive(s) designed to impart water absorption resistance to the board. Wax emulsions are commonly used for this purpose.

Wax. Refers to paraffin wax added usually in emulsified form to resin-bonded particleboard during the blending operation to impart a degree of water repellency to the product. See also Size.

Weathering Resistance. Resistance to degradation when exposed to the weather.

Wetability. A measure of the ability of a given surface to accept and spread liquids.

Wood. Xylem of fibrovascular tissues.

Wood Defects. A variety of generally degrading flaws which could occur during tree growth or afterward during manufacture.

Wood Flour. Very fine particles of wood resembling cereal flour.

Wood-Plastic Combination. Wood in which a liquid monomer has been entered and polymerized *in situ*.

Wood Wool. A type of excelsior; same as excelsior.

Working Life.   See Pot Life.

Yield.   The amount of wood material which actually ends up in the board as against the quantity used to produce that given amount.

Unit Conversions

These unit conversions appear in identical form in both Volumes 1 and 2.

LENGTH

FROM	TO	MULTIPLY BY
Micron	Centimeter	0.0001
Centimeter	Inches	0.3937
Meters	Feet	3.28083
Inches	Centimeter	2.5400

AREA

Square centimeters	Square inches	0.1550
Square inches	Square centimeters	6.4516
Square meter	Square feet	10.76
Square feet	Square meters	0.0929
Acres	Square feet	43560

VOLUME

Cubic centimeters	Cubic inches	0.0610
Cubic inches	Cubic centimeters	16.387
Cubic meters	Cubic feet	35.31
Cubic feet	Cubic meters	0.02832
Cubic meters	Cubic yards	1.308
Cubic yards	Cubic meters	0.7646

CAPACITY

Liters	Cubic feet	0.03531
Liters	British gallons	0.2200
Liters	U.S. gallons	0.26418
Cubic feet	Liters	28.317
U.S. gallons	Cubic feet	0.13368
Cubic feet	U.S. gallons	7.48052
British gallons	U.S. gallons	1.20094
British gallons	Cubic feet	0.1605
U.S. gallons	Liters	3.785
British gallons	Liters	4.546

WEIGHT

Grams	Ounces	0.03527
Ounces	Grams	28.350

Kilograms	Pounds	2.2046
Pounds	Kilograms	0.4536
Metric tons	Kilograms	1000
Metric tons	Short tons	1.10231
Metric tons	Long tons	0.98421
Metric tons	Barrels (376 pounds)	5.8633
Long tons	Short tons	1.12
Long tons	Barrels (376 pounds)	5.957
Short tons	Barrels (376 pounds)	5.3191

STRENGTH

Kilograms per square centimeter	Pounds per square inch	14.2233
Pounds per square inch	Kilograms per square centimeter	0.0703

TEMPERATURE

Degrees centigrade	Degrees Fahrenheit	$(°C. \times 1.8) + 32$

HEAT

Calories (kilogram)	Btu	3.96753

POWER

Kilowatts	Horsepower	1.34102
Horsepower	Kilowatts	0.746
Btu's per second	Horsepower	1.41460
Kilowatt hours	Btu's	3413.44

OTHERS

Pounds per inch	Kilograms per centimeter	0.1786
Kilograms per centimeter	Pounds per inch	5.600

Bibliography

This bibliography appears in identical form in both Volumes 1 and 2.

Aaron, J. R. 1966. The utilization of bark. Res. and Dev. Paper No. 32. U.K. For. Commission, London.

Abbott, E. T., and Firestone, F. A. 1933. Specifying surface quality. Mech. Eng. No. 55.

Adams, D. F., Monroe, F. L., Rasmussen, R. A., and Bamesberger, W. L. 1971. Volatile emissions from wood. Fifth Pclebd. Sympo. Proc., Wash. St. Univ., Pullman: 171–193.

Akers, L. E. 1966. Particleboard and hardboard. Pergamon Press, London.

Albers, K. 1971. Behavior of wood-based materials under transverse compression. Holz Roh-u Werkstoff 29(3): 94–96.

————. 1971. Measurement of modulus of rigidity of wood-based materials. Part 3: Modulus of rigidity of extruded boards. Holz Roh-u Werkstoff 29(12): 470–474.

Albrecht, J. W. 1968. The use of wax emulsions in particleboard production. Second Pclebd. Sympo. Proc., Wash. St. Univ., Pullman: 31–52.

Amthov, J. 1972. Paraffin dispersions for the waterproofing of particleboard. Holz Roh-u Werkstoff 30(11): 422–429.

Anderson, A. B., and Helge, K. 1959. Bark in hardboard. For. Prod. J. 9(4): 31A–35A.

————, and Punckel, W. J. 1950. Utilization of Douglas-fir bark in hardboard. Proc. For. Prod. Res. Soc., p. 301.

————, Breuer, R. J., and Nicholls, G. A. 1961. Bonding particleboards with bark extracts. For. Prod. J. 11(5): 226–227.

Anon. 1953. Fire tests on building materials and structures. British Stand. 476. Part 1. British Stand. Inst.

————. 1955. Surface roughness, waviness, and lay. Am. Standards Assn., B46–1.

————. 1955. Wood handbook. USDA Ag. Handbk. No. 72.

————. 1957. Fiberboard and particleboard. FAO, Rome.

————. 1957. New processes for molding parts with wood particles. System Collipress. Holz Roh-u Werkstoff 15(5): 229–230.

————. 1957. Wood waste utilization. 1954–56. Tech. Ser. Bull. No. 22. For. Prod. Res. Soc., Madison, Wis.

————. 1958. Standard method of test for impedance and absorption of acoustical materials by the tube method. Am. Soc. Test. and Mtls., Phila., Pa. Standard C384–58.

————. 1958. Standards for fire-retardant formulations P10–58. Am. Wood Pres. Assn., Manual of Recom. Practice.

————. 1959. A 3800-gallon tube of glue. Adhesives Age 2(9): 33.

————. 1959. Dimensional stability seminar. USDA Forest Service, For. Prod. Lab. Rept. No. 2145.

————. 1959. Fire test methods used in research at the Forest Products Laboratory. USDA Forest Service, For. Prod. Lab. Rept. No. R 1443 (Revised).

————. 1959. Handling liquid adhesives. Adhesives Age 2(12): 28.

————. 1959. Wood particle binder resins. Am. Cyanamid Co. Plastics and Resins Div. Tech. Data Bull. No. 2.

————. 1961. Commercial standards CS236–61 for mat-formed wood particle-board. USDC, Wash., D. C.

————. 1961. Standard E 69–50. Am. Soc. Test. and Mtls., Phila., Pa.

————. 1962. Bulk handling and storage of liquid synthetic resins. Martin-Marietta Corp., Tech. Service Publ. 2–1–1–1.

————. 1962. Cleaning of synthetic resin storage tanks. Martin-Marietta Corp., Tech. Service Publ. 2–1–1–3.

————. 1962. Handling and storage of chemicals for adhesives. Martin-Marietta Corp., Tech. Service Publ. 2–1–1–2.

————. 1965. A progressive program. Hermal Report, Munich, W. Germany, No. 17.

————. 1965. Better, cheaper, quicker rationalization tips. Hermal Report, Munich, W. Germany, No. 1.

————. 1965. Cooling of chipboard after press. Hermal Report, Munich, W. Germany, No. 2.

————. 1965. Plant without caul plates—just a fashion or something new? Hermal Report, Munich, W. Germany, No. 17.

————. 1965. Processing offers new opportunities for profit. Hermal Report, Munich, W. Germany, No. 17.

————. 1966. Book or ASTM standards. Part 16: Structural sandwich construction; wood; adhesives. Am. Soc. Test. and Mtls., Phila., Pa.

————. 1966. Complex shapes in molded wood components. F.D.M. 38(1): 73.

————. 1966. Inline mixing of UF resins and catalysts in particleboard manufacture. Borden, Inc., Adh. & Chem. Div., Ser. Bull. SB–110.

————. 1966. Molding furniture parts with Granulplast resins. Monsanto, Plastics Prod. & Resin Div., Tech. Bull. No. 6087.

————. 1966. Quality control in the particleboard industry. Borden, Inc., Adh. & Chem. Div., Ser. Bull. SB–108.

————. 1966. Rapid expansion of particleboard in the United States continues. Hermal Report, Munich, W. Germany, No. 17.

————. 1966. Resin application in particleboard manufacture. Borden, Inc., Adh. & Chem. Div., Ser. Bull. SB–107.

————. 1966. Symposium on mechanical barking of timber. FAO/ECE/LOG 162, 3 vols. FAO, Rome.

————. 1967. Board production opens in Australia. World Wood. December issue.

————. 1967. Bulk handling of liquid resin glue. Borden, Inc., Adh. & Chem. Div., Ser. Bull. SB–79c.

————. 1967. Evaluating properties of wood-base fiber and particle panel materials. Am. Soc. Test. and Mtls., Phila., Pa. Standard D1037–64.

————. 1967. Hardwoods: Important raw material for particleboard production. Hermal Report, Munich, W. Germany, No. 14.

————. 1967. Wood particleboard—What is it? Quest, Wash. St. Univ., Coll. of Eng., Pullman 5(2): 18.

————. 1968. Directory of debarking machines. FAO, Rome.

————. 1968. New caulless system teams with old line, triples board capacity. Wood & Wood Prod. February issue.

————. 1968. Tack in particleboard. Borden, Inc., Adh. & Chem. Div., Ser. Bull. SB–114.

————. 1969. Outside chip storage in B.C. interior. Research News, Vol. 12, No. 3, Dept. of Fish and Forestry, Canada.

————. 1969. Particleboard manufacture. Borden, Inc., Adh. & Chem. Div., Ser. Bull. SB–78i.

————. 1969. Particleboard plant emphasized versatility and quality control. Forest Industries. July issue.

————. 1969. Sees particleboard consumption at 50 pounds per capita in five years. Wood & Wood. Prod. 70(7): 41–42.

————. 1970. Accelerated aging of phenolic resin-bonded particleboard. For. Prod. J. 20(10): 26–27.

————. 1970. New plant turns out $8\frac{1}{2}$ × 42 ft. particleboard; features cut-to-size. Wood & Wood Prod. June issue.

————. n.d. Bison particleboard plants. Pamph., Bahre Mettallwerk KG, Springe/Hannover, W. Germany.

————. n.d. From log to chipboard. Pamph., Carl Schenck Maschinenfabrik GmbH, Darmstadt, W. Germany.

————. n.d. Granulated wood molding. Monsanto Chem. Co., Plastics Div., Info. Bull. No. 1027.

————. n.d. Induma particleboard plants. Pamph., Induma Industrie Maschinen GmbH, Hamburg, W. Germany.

Arola, R. A. 1973. Compression debarked chips from a whole-tree chipper. USDA Forest Service, Res. Note NC–147, 4 pp.

————, and Erickson, J. R. 1973. Compression debarking of wood chips. USDA Forest Service, Res. Paper NC–85, 11 pp.

Arsenault, R. D. 1964. Fire retardant particleboard from treated flakes. For. Prod. J. 14(1): 33–39.

Auchter, R. J. 1973. Recycling forest products retrieved from urban waste. For. Prod. J. 23(2): 12–16.

Back, E. L., Didriksson, E. I. E., Johanson, F., and Norberg, G. K. 1971. The hot, dry mouldability of hardboard. For. Prod. J. 21(8): 96–101.

Badanoiu, G. 1970. Experimental utilization of the wood of various species in the production of fiberboards. Industr. Lemn. 21(7): 246–253.

———. 1971. Wood particleboards made exclusively from wood wastes in small capacity installations. Industr. Lemn. 22(5): 159–166.

Baer, E. (ed.). 1964. Engineering design for plastics. Reinhold, New York.

Balint, J. E. 1960. A method for proportioning resin and wax in particleboard. For. Prod. J. 10(11): 567–570.

Barrington, B. C. 1967. The Ortmann flaker. First Pclebd. Sympo. Proc., Wash. St. Univ., Pullman: 95–107.

Bauza, J. 1971. Utilization of Scots pine bark in particleboard production. Prace Inst. Tech., Drewna 18(1): 3–23.

Baxter, G. F., and Kreibich, R. E. 1973. A fast curing phenolic adhesive system. For. Prod. J. 23(1): 17–22.

Becker, G. 1969. Tests on wood chipboards with *Hylotrupes bajulus* larvae and termites. Mater. U. Organ., Beih. No. 2: 27–41.

———, and Deppe, H. J. 1970. Behavior of untreated particleboard and particleboard chemically treated against organisms. USDA Transl. FPL–711, pp. 13.

Behr, E. A. 1972. Decay and termite resistance of medium-density fiberboards made from wood residue. For. Prod. J. 22(12): 48–51.

Bender, F. 1959. Spruce and balsam bark as a source of fiber products. Pulp & Paper Mag. Canada 60(9): T275–T278.

Bennett, H. 1963. Industrial waxes. Vol. 1. Chemical Publishing Co., New York.

Berchem, T. E. 1970. The effect of particle size on strength, dimensional properties and surface roughness of hickory particleboard. M.S. thesis, South. Ill. Univ., Carbondale.

Berger, V. 1964. Urea binders with low content of free formaldehyde. Holztechnol. 4(5): 79.

Berrong, H. B. 1968. Efficiency in sizing of particleboard. Second Pclebd. Sympo. Proc., Wash. St. Univ., Pullman: 55–74.

Bersenev, A. P. 1970. Some methods of increasing the durability of particleboards. Lesn. Z. 13(1): 114–116.

Bhagwat, S. 1971. Physical and mechanical variations in cottonwood and hickory flakeboards made from flakes of three sizes. For. Prod. J. 21(9): 101–103.

Bibby, R. D. 1956. Manufacture and use of particleboard. For. Prod. J. 6(5): 169–172.

Biblis, E. J., and Lo, C. F. 1968. Effect on setting of southern pine cement mixtures. For. Prod. J. 18(8): 28–34.

Binder, K. A. 1967. Resin application and quality in particleboard manufacture. First Pclebd. Sympo. Proc., Wash. St. Univ., Pullman: 235–250.

Birks, A. S. 1972. Particleboard blow detector. For. Prod. J. 22(6): 23–26.

Blais, J. F. 1959. Amino resins. Reinhold, New York.

Blanks, R. F., and Kennedy, H. L. 1955. The technology of cement and concrete. Wiley, New York.

Bockhold, L. W. 1971. Filtering mediums in particleboard air collection systems. Fifth Pclebd. Sympo. Proc., Wash. St. Univ., Pullman: 263–274.

Bokhovkin, I. M., Evseen, G. A., and Kovalenko, N. P. 1971. Effect of water-soluble organic substances on the processes of cement hardening. Lesn. Z. 14(2): 104–108.

Bollerslev, K. 1958. Bark processing problems. For. Prod. J. 18(6): 19–20.

Bosch, J. M. 1969. Experiences with closed circuit television in a particleboard plant. Third Pclebd. Sympo. Proc., Wash. St. Univ., Pullman: 145–148.

Bosshard, H. H., and Futo, L. P. 1963. Specific staining for detection of gluing with urea and phenolic resins in plywood. Holz Roh-u Werkstoff 21(6): 225–228.

Branion, R. 1961. Fiberboards from bark-wood mixtures. Pulp & Paper Mag. Canada 62(11): T506–T508.

Brenden, J. J. 1971. Usefulness of a new method for measuring smoke yield from wood species and panel products. For. Prod. J. 21(12): 22–28.

Bridges, R. R. 1971. A quantitative study of some factors affecting the abrasiveness of particleboard. For. Prod. J. 21(11): 39–41.

Briggs, N. J. 1958. Frequency distribution; an aid to control. For. Prod. J. 8(11): 34A–35A.

Brockschmidt, K. H. 1960. Fundamentals of molding wood particles. For. Prod. J. 10(4): 179–183.

Brown, G. E., and Alden, H. M. 1960. Protection from termites: Penta for particleboard. For. Prod. J. 10(9): 434–438.

Brown, W. H. 1971. Particleboard in building: A guide to its manufacture and use. Timb. Res. and Develop. Assn., Hughenden Valley, High Wycombe, Bucks, England, 43 pp.

Bruce, Richard, W. 1970. The economics of marketing particleboard 1970. Fourth Pclebd. Sympo. Proc., Wash. St. Univ., Pullman: 47–76.

Brumbaugh, J. I. 1960. Effect of resin spot size on particleboard properties. Unpublished research, Inst. of Tech., Wash. St. Univ., Pullman.

———. 1960. Effect of flake dimensions on properties of particleboard. For. Prod. J. 10(5): 243–246.

———. 1967. Overlaying of particleboard. First Pclebd. Sympo. Proc., Wash. St. Univ., Pullman: 431–446.

———. 1969. Plant experiences in controlling moisture. Third Pclebd. Sympo. Proc., Wash. St. Univ., Pullman: 61–72.

Brunauer, S., and Copeland, L. E. 1964. The chemistry of concrete. Sci. Am. 210: 30–88.

Bryan, E. L. 1971. Modern Management techniques for particleboard operations. Fifth Pclebd. Sympo. Proc., Wash. St. Univ., Pullman: 133–137.

———. 1962. Dimensional stability of particleboard. For. Prod. J. 12(12): 572–576.

Bryant, B. S. 1958. Commercial Standards and product specifications. For. Prod. J. 8(11): 31A–32A.

Bryant, L. H., and Humphreys, F. R. 1958. Building boards with sawmill waste. Composite Wood 5(2–3): 41–46.

Buikat, E. R. 1971. Operating problems with dryers and potential solutions. Fifth
    Pclebd. Sympo. Proc., Wash. St. Univ., Pullman: 209–216.
Buntrock, K. 1971. Scrubbers in particleboard plants. Fifth Pclebd. Sympo. Proc.,
    Wash. St. Univ., Pullman: 301–316.
Burrows, C. H. 1958. Floor tiles from planer shavings. For. Prod. Res. Lab., Oreg.
    St. Univ., Corvallis, Info. Cir. No. 12.
———. 1959. Floor tile from Douglas-fir bark. For. Prod. Res. Lab., Oreg. St.
    Univ., Corvallis, Info. Cir. No. 13.
———. 1960. Particleboard from Douglas-fir bark—without additives. For. Prod.
    Res. Lab., Oreg. St. Univ., Corvallis, Info. Cir. No. 15.
———. 1961. Some factors affecting resin efficiency in flakeboard. For. Prod. J.
    11(1): 27–33.
Buschbeck, L., and Kehr, E. 1960. Investigations of shortening of compression
    time in hot compression of particleboards. Holztechnol. 1(1): 112–123.
———, ———, and Jensen, U. 1961. Investigations on the suitability of various
    wood types and varieties for the manufacture of particleboards. Part 1: Red
    beech and spruce. Holztechnol. 2(1): 99–110.
———, ———, Scherfke, R., and Jensen, U. 1965. Investigation on the suitability
    of various wood species and assortments for chipboard manufacture. Part 1:
    Red beech and pine. USDA Transl. FPL–626.
———, ———, ———, and ———. 1965. Investigation on the suitability of var-
    ious wood species and assortments for chipboard manufacture. Part 3: Pine
    brushwood. USDA Transl. FPL–627.
Cagle, C. V. 1968. Adhesive bonding: Techniques and applications. McGraw-Hill,
    New York.
Carre, J. 1970. Effect of the type and shape of raw material on the physical and
    mechanical properties of particleboard. Rapp. Sta. Tech. For., Gembloux,
    pp. 267.
Carroll, M. N. 1963. Efficiency of urea- and phenol-formaldehyde in particleboard.
    For. Prod. J. 13(3): 113–120.
———. 1963. Whole wood and mixed species as raw material for particleboard.
    Inst. of Tech., Wash. St. Univ., Pullman, Bull. 274, 45 pp.
———. 1970. Relationship between driving torque and screwholding strength in
    particleboard and plywood. For. Prod. J. 20(3): 24–30.
———. 1972. Measuring screw withdrawal with a torque wrench. For. Prod. J.
    22(8): 42–46.
———, and McVey, D. 1962. An analysis of resin efficiency in particleboard.
    For. Prod. J. 12(7): 305–310.
———, and Wrangham, B. 1970. The measurement of surface strength in particle-
    board. For. Prod. J. 20(7): 28–33.
Carruthers, J. F. S. 1959. Heat penetration in the pressing of plywood. For. Prod.
    Res. Bull. No. 44, Her Majesty's Stationery Office, London.
Carter, R. K. 1969. Particleboard production by the caulless process. Paper pre-
    sented at 20th annual meeting in San Francisco. For. Prod. Res. Soc., Madison,
    Wis.
———. 1969. Weighing devices and instruments used in particleboard manufacture.
    Third Pclebd. Sympo. Proc., Wash. St. Univ., Pullman: 103–112.

Caughey, R. A. 1955. Development and market potential of particleboard. For. Prod. J. 5(8): 19A–20A.

Chang, Y. 1954. Anatomy of common North American pulpwood barks. Tappi mono. No. 14.

———. 1954. Bark structure of North American conifers. USDA Tech. Bull. No. 1095.

———, and Mitchell, R. L. 1955. Chemical composition of common North American pulpwood barks. Tappi 38(5): 315–320.

Charney, M. 1970. Protection systems for dust explosions. Fourth Pclebd. Sympo. Proc., Wash. St. Univ., Pullman: 197–208.

Chen, T. Y., Paulitsch, M., and Soto, G. 1972. On the suitability of the aerial biomass from spruce thinnings as raw material for particleboard. Holz Roh-u Werkstoff 30(1): 15–18.

Chow, P. 1971. Dimensional stability of particleboard in reaction to humidity changes. Furniture Design and Mfg. September issue.

———. 1972. Modulus of elasticity and shear deflection of walnut-veneered-particleboard composite beams in flexure. For. Prod. J. 22(11): 33–38.

———, Walters, C. S., and Guiher, J. K. 1971. pH measurements for pressure-refined plant-fiber residues. For. Prod. J. 21(12): 51–52.

———, ———, and ———. 1973. Specific gravity bulk density and screen analysis of midwestern plant-fiber residues. For. Prod. J. 23(2): 57–60.

Chow, S.Z., and Mukai, H.N. 1969. An infrared method of determining phenolformaldehyde resin content in fiber and wood product. For. Prod. J. 19(5): 57–60.

Christensen, L. E., and Lyneis, R. G. 1948. The effect of sugar and wood extracts on the properties of portland cement mixtures. B.S. thesis, Univ. of Wis., Madison.

Christensen, R. L. 1972. Test for measuring formaldehyde emission from formaldehyde resin-bonded particleboards and pulpwood. For. Prod. J. 22(4): 17–20.

Chugg, W. A., Wood, T., and James, P. E. 1965. The gluability of hardwoods for structural purposes. Timb. Res. and Develop. Assn., Res. Rept. C/RR/22, England, 54 pp.

Clad, W. 1967. Phenolic formaldehyde condensates as adhesives for particleboard manufacture. Holz Roh-u Werkstoff. 25(4): 137–147.

———. 1970. On the use of waste paper for particleboard production. USDA Transl. FPL–713.

———. 1971. Rubber wastes for particleboards. Holz-Zbl. 97(85): 1235.

———. 1972. Trend analysis of adhesives for wood base panel products. Holz Roh-u Werkstoff 30(4): 127–129.

———, and Pommer, E. H. 1971. Open-air weathering tests on particleboard. Part 1: The effect of resin content on durability. Holz-Zbl. 97(28): 397–398.

———, and ———. 1971. Open-air weathering tests on particleboard. Part 2: Durability of particleboards glued with melamine and phenolic resins. Holz-Zbl. 97(32): 453–454.

———, and ———. 1971. Open-air weathering tests on particleboard. Part 3: Resistance to fungi of particleboards in exterior use. Holz-Zbl. 97(35): 505–506.

Clare, K. E. 1956. Further studies on the effect of organic matter on the setting of soil mixtures. J. Appl. Chem. 6: 317–324.

———, and Sherwood, P. T. 1954. The effect of organic matter on the setting of soil cement mixtures. J. Appl. Chem. 4: 625–630.

Clark, J. d'A. 1955. A new dry-process multi-ply board. For. Prod. J. 5(8): 209–213.

Clarke, L. N. 1954. A method of measuring the thermal conductivity of poor conductors. Aust. J. Appl. Sci. 5(2): 178–182.

Clermont, L. P., and Schwartz, H. 1948. Studies on the chemical composition of bark and its utilization for structural boards. Proc. For. Prod. Res. Soc., p. 130.

Commercial Standard CS236–66. 1966. Mat-formed particleboard. USDC, Wash. D.C.

Connelly, T. J. 1955. The Kreibaum process of extruded core board. For. Prod. J. 5(8): 21A–23A.

Cooke, W. H., and Frashour, R. G. 1955. Resin application in attrition-mill type particleboard. For. Prod. J. 5(8): 214–218.

Crafton, J. M. 1956. Extruded chip core board. For. Prod. J. 6(5): 173–175.

Cranch, R. C. 1964. What wood can you use in particleboard. Wood Res., Timb. Eng. Co., Wash., D.C. 34(4): 2–7.

Crawford, R. J. 1967. Pressing techniques, problems, and variables. First Pclebd. Sympo. Proc., Wash. St. Univ., Pullman: 349–356.

Cunderlik, O., and Rajkovic, E. 1970. Some problems of the occurance of thickness variations in the pressing of particleboards. Part 1: Technological factors influencing the thickness of particleboards after pressing. Drev. Vyskum 1: 21–33.

Currier, R. A. 1962. Compression of Douglas-fir plywood in various hot pressing cycles. For. Prod. Res. Lab., Oreg. St. Univ., Corvallis, Info. Cir. No. 17.

Curry, W. T. 1957. The strength properties of plywood. Part 3: The influence of the adhesive. For. Prod. Res. Bull. No. 39. Her Majesty's Stationery Office, London.

Curtis, F. W. 1944. High frequency induction heating. McGraw-Hill, New York.

Curtu, I., and Fleishcer, H. 1970. Elastic constants of particleboards and fiberboards produced in Romania. Industr. Lemn. 21(3): 92–98.

Davis, T. C. 1966. Effect of blue stain on setting of excelsior cement mixtures. For. Prod. J. 16(6): 49–50.

Delmonte, J. 1947. The technology of adhesives. Reinhold, New York. 355 pp.

Deppe, H. J. 1969. Developments in the production of multi-layer foamed wood particleboard. For. Prod. J. 19(7): 27–33.

———. 1971. Protection of isocyanate-bonded particleboards. Holz Roh-u. Werkstoff 29(6): 217–218.

———, and Ernst, K. 1964. Technologie der Spanplatten. Holz-Zbl., pp. 156–177.

———, and ———. 1966. On reducing compacting time in chipboard manufacture. USDC Transl. FPL–670.

———, and ———. 1971. Isocyanates as adhesives for particleboards. Holz Roh-u Werkstoff 29(2): 45–50.

———, and Gersonde, M. 1969. Production and testing of preserved wood chipboard. Mater. u. Organ., Beih. No. 2: 123–136.

————, and Hoffmann, A. 1971. Conifer bark for particleboards. Holz-Zbl. 97(152): 2195–2197.

————, and Stolzenburg, R. 1971. Coated particleboard for exterior use. Holz-Zbl. 97(19): 247–248.

————, and ————. 1971. Coated particleboard for exterior use. Holz-Zbl. 97(25): 349–350.

Dmitrieva, G. A. 1972. Logging waste—a valuable raw material for the manufacture of fiberboards. Derev. Prom. 21(1): 6.

Donohue, F. 1961. Granulated wood molding. Industrial woodworking 7, Nos. 6, 7.

————. n.d. Molded furniture parts. Monsanto. Chem. Co., Plastics Div., Springfield, Mass.

Dost, W. A. 1971. Redwood bark fiber in particleboard. For. Prod. J. 21(10): 38–43.

Draganov, S. M. 1968. Fire retardants in particleboard. Second Pclebd. Sympo. Proc., Wash. St. Univ., Pullman: 75–121.

Eby, L. T., and Brown, H. P. 1969. Thermosetting adhesives. In Treat. on adhesion and adhesives, ed. R. L. Patrick. Marcel Dekker, New York, pp. 77–171.

Ehlers, W. 1958. Uber die Bestimmung der Gute von Holzoberflachen Holz Roh-u Werkstoff 16: 49.

Eibenstein, A. 1971. Directed, controlled glue application to chips. The Teutomatic multi-stage system. Holz-Zbl. 97(151): 2186.

Eickner, H. W. 1942. Gluing characteristics of fifteen species of wood with cold setting urea resin glue. USDA Forest Service, For. Prod. Lab., FPL mimeo No. 1342, Madison, Wis.

Eisner, K., and Kolejack, M. 1958. Some basic problems concerning the production of chipboards in short pressing cycle. Drevarsky Vyskum 3(2): 223–242.

Elbert, A. A. 1962. A heat treatment method for improving the water repellency of chipboards. Derev. Prom. 11(10): 3–6.

————. 1970. Moisture resistance of particleboards. Izdatelstvo Lesnaja Prommyslennost, Moscow. pp. 97.

Elliot, D. R., and deMontmorency, W. H. 1963. The transportation of pulpwood chips in pipelines. Pulp & Paper Res. Inst. Canada, Montreal. Walds. Res. Index, No. 144.

Ellwood, E. L. 1954. Properties of American beech in tension and compression perpendicular to the grain and their relation to drying. School of For., Yale Univ., New Haven, Bull. No. 61.

Elmendorf, A. 1967. New tri-panel combines veneer core and wood strands in product that utilizes entire log. Plywd. & Panel, February issue.

Ely, H. M. 1970. Fire and explosion prevention, detection, and extinguishing. Fourth Pclebd. Sympo. Proc., Wash. St. Univ. Pullmann: 209–215.

Endicott, L. E., and Frost, T. R. 1967. Correlations of accelerated and long-term stability tests for wood-based composite products. For. Prod. J. 17(10): 35–40.

Erickson, J. R. 1971. Bark-chip segregation: A key to whole-tree utilization. For. Prod. J. 21(9): 110–113.

Ernsberger, F. M., and France, W. G. 1945. Portland cement dispersion by absorption of calcium-lignosulfate. Ind. & Eng. Chem. 37: 598–600.

Ettling, B. F., and Adams, M. F. 1966. Quantitative determination of phenolic resins in particleboard. For. Prod. J. 16(6): 25–28.

Evseev, G. A. 1970. Speeding up the hardening processes for light concrete based on cement and wood waste. Les. Z. 13(2): 162–165.

Fahey, D. J., and Pierce, D. S. 1971. Resistance of resin-impregnated paper overlays to accelerated weathering. For. Prod. J. 21(11): 30–38.

Fahrni, F. 1956. Essential features of batch formation for particleboard manufacture. FAO/ECE/Board Cons. Paper 5–31.

Farber, E. Chemicals from bark. For. Prod. J. 9(4): 25A–27A.

Fehn, H. 1954. The manufacture of chipboard—a technological and industrial evaluation. Council of Sci. and Ind. Res., Australia. Transl. No. 2615.

Feist, W. C., Springer, E. L., and Hajny, G. L. 1971. Encasing wood chip piles in plastic membranes. Tappi 54(7): 1140–1142.

Ferguson, I. S. 1972. Wood chips and regional development. Australian Forestry 36(1): 15–23.

Fickler, H. H. 1965. Trends in the pressing technology of particleboards. Holzindustrie 18(4): 95–96.

———, and Lundmaker, L. E. 1966. Experience gathered on cation-active wax dispersions. Holz Roh-u Werkstoff 24(7): 290–294.

Fischbein, W. 1957. Continuous press manufacture. FAO/ECE/Board Cons. Paper 5–35.

Fischer, A., and Merkle, G. 1963. Testing of particleboard surfaces. Holz Roh-u Werkstoff 21(3): 96–104.

Fischer, R. 1957. Conditioning particleboards. FAO/ECE/Board Cons. Paper 5–37.

Fisken, A. M. 1956. Essential features of continuous formation for particleboard manufacture. FAO/ECE/Board Cons. Paper 5–32.

Fleming, H. 1962. Factors affecting the formation and properties of material with fiberous structure. Part 2. Holztechnologie 3(2): 105–110.

———. 1964. Factors affecting the formation and properties of material with fiberous structure. Parts 1 and 2. USDA Transl. FPL–602.

Foster, W. G. 1967. Species variation. First Pclebd. Sympo. Proc., Wash. St. Univ., Pullman: 13–20.

Frashour, R. G. 1967. The Columbia microfelter. First Pclebd. Sympo. Proc., Wash. St. Univ., Pullman: 319–325.

———. 1967. The Forrest Industries blender. First Pclebd. Sympo. Proc., Wash. St. Univ., Pullman: 301–307.

Freeman, H. A. 1959. Relation between physical and chemical properties of wood and adhesion. For. Prod. J. 9(12): 451–458.

Galligan, A. F. 1969. Recent glues and gluing research applied to particleboard. For. Prod. J. 19(1): 44–51.

Galper, G. E., and Tsypkina, M. N. 1970. Changes in certain physical characteristics of wood and its constituents during the manufacture of fiberboards. Khim. Drev. No. 5, pp. 151–162.

Gamov, V. V. 1970. The structural and mechanical properties of particleboards. Lesn. Z. 13(3): 81–85.

————. 1971. Effect of particle width on the strength of particleboards in tension parallel to surface. Lesn. Z. 14(2): 73–77.

Gatchell, C. J., and Heebink, B. G. 1964. Effect of particle geometry on properties of molded wood-resin blends. For. Prod. J. 14(11): 501–506.

————, ————, and Hefty, F. V. 1966. Influence of component variables on properties of particleboard for exterior use. For. Prod. J. 16(4): 46–59.

Gerard, J. C. 1966. Profiles in dimensional behavior of particles in simulated particleboard construction. For. Prod. J. 16(6): 40–48.

Gerlachov, T. 1970. Influences of the presence of NaOH in the chip mixture on particleboard properties. Drevo. 25(1): 11–12, 16.

————. 1970. Variation of the properties of chip mouldings in relation to the type of phenolic resin used. Drev. Vyskum 3: 123–143.

Gertjejansen, R. O., and Haygreen, J. G. 1971. Torsion shear test for particleboard adapted to a universal testing machine. For. Prod. J. 21(11): 59–60.

Ginzel, W. 1971. On the chemical-hydrolytic degradation of urea-formaldehyde resin in particleboard after three years weathering. Holz Roh-u Werkstoff. 29(8): 301–305.

————. 1973. Contribution to the hydrolysis of urea-resin-bonded particleboard. Holz Roh-u Werkstoff 31(1): 18–24.

Gluck, D. G. 1971. Selection and field testing of coatings for exterior hardboard products. For. Prod. J. 21(11): 24–29.

Godshall, W. D., and Davis, J. H. 1969. Acoustical absorption properties of woodbase panel materials. USDA Forest Service, For. Prod. Lab. Res. Pap. FPL–104.

Goldstein, I. S., and Drecher, W. A. 1961. A non-hygroscopic fire retardant treatment for wood. For Prod. J. 11(5): 235–237.

Golubov, I. A., and Panov, V. V. 1971. Nomogram for determining the content of binder in the chip-glue mix. Derev. Prom. 20(9): 4–5.

Gomonov, A., and Levin, I. 1970. Arbolit (wood/concrete) made with belite-slurry cement. Lesn. Prom. 9: 14–15.

Graf, G., Koch, H., Schiene, R., and Schuster, E. 1972. Producing boards from ammonia-plasticized fibers and their strength. Part 2. Holztechnol. 13(3): 152–155.

Graham, R. F., and Parrish, F. 1961. The economic possibilities of wood particleboard manufacture in Maine. Div. Res. & Plan., Dept. Econ. Devel., State House, Augusta.

Graho, J., Jr., and Williams, F. H. 1954. The manufacture of molded articles from resin-bonded granulated wood. For. Prod. J. 4(9): 228–229.

Grant, E. L. 1952. Statistical quality control. 2nd ed. McGraw-Hill, New York.

Graser, M. 1962. Temperature variation in industrially manufactured particleboard. Holz-Zbl. Supplement, pp. 88, 137.

Gressel, P. 1969. Investigations on weather aging of particleboard. A comparison between two-year weathering and three different accelerated test methods. USDA Transl. FPL–692.

————. 1972. The effect of time, climate, and loading on the bending behavior of woodbase materials. Part 1: Previous investigations, testing plan, research methods. Holz. Roh-u Werkstoff 30(7): 259–266.

———. 1972. Effect of climate and loading on the creep behavior of woodbase materials. Part 2: Test results in dependency on the creep parameters. Holz Roh-u Werkstoff 30(9): 347–355.

———. 1972. Effect of climate and loading on the creep behavior of woodbase material. Part 3: Discussion of results. Holz Roh-u Werkstoff 30(12): 479–488.

Grigor, A., Fleischer, H., and Mitisor, A. 1968. The static bending strength of particleboard. Holz Roh-u Werkstoff 26(8): 287–290.

Gumbrecht, D. L. 1969. "Tailoring" phenolic adhesives. For. Prod. J. 19(6): 38–40.

Gunn, J. M. 1963. Wafer dimension control; number one design criteria for plant producing particleboard for building construction. For. Prod. J. 13(4): 163–167.

Gunther, B., Luthardt, H., and Nicht, J. 1971. Conversion of sawdust and bark into board products. Part 2. Holzindustrie 24(8): 231–236.

Haas, H. G., and Munk, E. E. 1967. Molded wood particles open new product areas. Wood & Wood Prod. November issue.

Hadley, D. 1959. Will particleboard replace wood cores. Veneers & Plywd. 53(11): 18–21.

Hagenstein, P. R. 1964. The location decision for wood using industries in the northern Appalachians. USDA Forest Service, NEFES, Upper Darby, Pa., USFS Res. Paper NE-16. 36 pp.

Haigh, A. H., Jr. 1967. The Bison blenders. First Pclebd. Sympo. Proc., Wash. St. Univ., Pullman: 275–289.

Hailie, D. A. 1957. Quality control in sanding particleboard. For. Prod. J. 7(7): 85.

Hall, R. B., Leonard, J. H., and Nicholls, G. A. 1960. Bonding particleboard with bark extracts. For. Prod. J. 10(5): 263–272.

Halligan, A. F. 1969. Recent glues and gluing research applied to particleboard. For. Prod. J. 19(1): 44–51.

———. 1970. A review of thickness swelling in particleboard. Wood Sci. Tech. 4(4): 301–312.

———, and Schniewind, A. P. 1972. Effect of moisture on physical and creep properties of particleboard. For. Prod. J. 22(4): 41–48.

Hamilton, J. F. 1961. The end use approach to particleboard standards. For. Prod. J. 11(5): 234.

Hann, R. A., Black, J. M., and Blomquist, R. F. 1962. How durable is particleboard. For. Prod. J. 12(12): 577–584.

———, ———, and ———. 1963. How durable is particleboard. Part 2: The effect of temperature and humidity. For. Prod. J. 13(5): 169–174.

Harkin, J. M., and Row, J. W. 1969. Bark and its possible uses. USDA Forest Service, For. Prod. Lab. Res. Note FPL-091.

Harris, E. E., and Stamm, A. J. 1953. Chemical processing of wood. Chem. Publ. Co., New York.

Harrison, C. W. 1971. Handling air and keeping it clean. Fifth Pclebd. Sympo. Proc., Wash. St. Univ., Pullman: 235–242.

Hatton, J. V. 1970. Precise studies on the effect of outside chip storage on fiber yield: white spruce and lodgepole pine. Tappi 53(4): 627–38.

Hart, C. A., and Rice, J. T. 1963. Some observations on the development of a laboratory flakeboard process. For. Prod. J. 13(11): 483–488.

Haygreen, J. G., and French, D. W. 1971. Some characteristics of particleboards from four tropical hardwoods of Central America. For. Prod. J. 21(2): 30–33.

———, and Gertjejansen, R. O. 1971. Improving the properties of particleboard by treating the particles with phenolic impregnating resin. Wood & Fiber 3(2): 95–105.

———, and ———. 1972. Influence of the amount and type of phenolic resin on the properties of a wafer-type particleboard. For. Prod. J. 22(12): 30–34.

———, and Turkia, K. 1972. Manufactured housing systems and their relation to the use of wood products. For. Prod. J. 22(8): 13–16.

Heape, G. C. 1970. Computer analysis for cut-to-size production control. Fourth Pclebd. Sympo. Proc., Wash. St. Univ., Pullman: 77–95.

Heebink, B. G. 1960. A new technique for evaluating show-through of particleboard cores. For. Prod. J. 8(7): 379–388.

———. 1960. Exploring molded particleboard. Wood & Wood Prod. 65(5): 36.

———. 1964. Improving planer residue for better utilization. Hitchcock's Woodworking Digest 66(7): 40–42.

———. 1967. A look at degradation in particleboards for exterior use. For. Prod. J. 17(1): 59–66.

———. 1967. A procedure for quickly evaluating dimensional stability of particleboard. For. Prod. J. 17(9): 77–80.

———. 1967. Wax in particleboards. First Pclebd. Sympo. Proc., Wash. St. Univ. Pullman: 251–268.

———. 1972. Irreversible dimensional changes in panel materials. For. Prod. J. 22(5): 44–48.

———, and Hann, R. A. 1959. Stability and strength of oak particleboards. For. Prod. J. 9(7): 236–242.

———, and Haskell, H. H. 1962. Effect of heat and humidity on the properties of high-pressure laminates. For. Prod. J. 12(11): 542–548.

———, and Hefty, F. V. 1968. Steam post treatments to reduce thickness swelling of particleboard. USDA Forest Service, For. Prod. Lab. Rept. FPL–0187.

———, and Lewis, W. C. 1967. Thick particleboards with pulp chip cores—possibilities as roof decking. USDA Forest Service, For. Prod. Lab. Rept. FPL–0174.

———, Hann, R. A., and Haskell, H. H. 1964. Particleboard quality as affected by planer shaving geometry. For. Prod. J. 14(10): 486–494.

———, Kuenzi, E. W., and Maki, A. C. 1964. Linear movement of plywood and flakeboards as related to the longitudinal movement of wood. USDA Forest Service, For. Prod. Lab. Res. Note 073.

Hemming, C. B. 1962. Wood gluing. In Handbook of adhesives, ed. I. Skeist. Reinhold, New York.

Henker, W. 1960. The use of continuous conveyors in automatic chipboard plants.

House Journal No. 8, Carl Schenck Maschinenfabrik GmbH, Darmstadt, W. Germany.

Herdey, O. 1956. Synthetic resins and other additives. FAO/ECE/Board Cons. Paper 4–18.

Heritage, C. C. 1957. A research program to determine the advisability of engaging in particleboard business. For. Prod. J. 7(11): 20A–23A.

Herrick, F. W., and Conca, R. J. 1960. The use of bark extracts in cold-setting, water-proof adhesives. For. Prod. J. 10(7): 361–368.

Hesch, R. 1970. A comparison of single and multiple presses: Technology and economic efficiency of particleboard manufacture as affected by press systems. Holz-Zbl. 96(66): 1004–1006.

Higgins, H. G. 1946. A critical temperature range in the plastic deformation of plywood. J. Council Sci. and Ind. Res., Australia 19: 455–462.

Hinselmann, D., Jentsch, S., and Drechsler, W. 1965. Investigations of manufacturing medium-density fiberboard according to the fiber drying process. USDA Transl. FPL–630.

Hopkins, B. R. 1970. High frequency curing of fiberboard. Fourth Pclebd. Sympo. Proc., Wash. St. Univ., Pullman: 253–279.

Horner, K. M. 1968. Plant experience and problem with cutting and sanding. Second Pclebd. Sympo. Proc., Wash. St. Univ., Pullman: 189–200.

———. 1969. Shipping particleboard. Third Pclebd. Sympo. Proc., Wash. St. Univ., Pullman: 193–214.

Hse, C. Y. 1971. Wettability of southern pine veneer by phenol-formaldehyde wood adhesives. For. Prod. J. 22(1): 51–56.

Huber, H. A. 1958. Preservation of particleboards and hardboard with pentachlorophenol. For. Prod. J. 8(12): 357–360.

Huffaker, E. M. 1962. Use of planer will residues in wood-fiber concrete. For. Prod. J. 12(7): 298–301.

Hunt, D. G. 1970. Measurements of elastic and creep properties of particleboard in tension and compression. Wood Sci. 2(4): 212–220.

Hunt, G. M., Garrett, G. A. 1967. Wood preservation. 3rd ed. McGraw-Hill, New York.

Hunt, M. O. 1962. Particleboard industry: Facts and references. Mimeo F–42, Coop. Ext. Serv., Purdue Univ., Lafayette, Ind.

———. 1970. The prediction of the elastic constants of particleboard by means of a structural analogy. Ph.D. diss., N.C. St. Univ., Raleigh.

Hyde, P. E., and Corder, S. E. 1971. Transportation costs in Oregon for wood and bark residues. For. Prod. J. 21(10): 17–25.

Isaacson, D. E. 1958. Statistical quality control in particleboard production. For. Prod. J. 8(11): 30A–31A.

Istrate, V., Filipescu, G. H., and Stefu, C. 1965. The influence of chip dimension on the consumption of glue and on the quality of particleboard. USDC Transl. FPL–613.

Iwashita, M., Ishihara, S., and Matsuda, T. 1963. Studies on particleboard. Part 4: Suitability of species as raw material for particleboard production (No. 1). Manufacture of particleboard using wooden extractives for particle bonding

Part 1. Bull. Gov. For. Exp. Station. No. 148. Tokyo Univ., Meguro: 73–92.

———, Matsuda, T., and Ishihara, S. 1960. Studies on particleboard. Part 3: Studies on the pressing. Part 1: On the curing condition, specially moisture content of wood particles. Bull. Gov. For. Exp. Station. No. 126. Tokyo Univ., Meguro: 63–89.

Jackson, D., and Savory, J. G. 1968. The decay resistance of wood-fiber building boards and particleboards. Int. Biodeterior. Bull. 4(2): 83–88.

Jaudon, J. L. 1970. Wood as raw material for particleboards. Rev. for. franc. 22(6): 587–602.

Jayne, B. A. 1966. Models and analyses in the physics of fiberous materials. For. Prod. J. 16(6): 51–59.

Jensen, U., and Kehr, E. 1970. Investigations on processing fines in particleboard manufacture. Part 1: Investigations and derivations for definition of fines and dust. Holztechnol. 11(2): 97–100.

———, and ———. 1971. Investigations on processing fines in particleboard manufacture. Part 2: Addition of sander dust and fines to the surface layer and core of particleboards. Holztechnol. 12(1): 3–8.

Jensen, W. 1966. The chemical utilization of bark. Sympo. Mech. barking of timber FAO/ECE/LOG/162. Three vols. FAO, Rome.

———, Tremer, K. E., Sierila, R., and Wartiovaara, V. 1963. The chemistry of bark. In the chemistry of wood, ed. B. L. Browning. Wiley, New York: 587–666.

Jenuleson, W. R. 1962. Requirements for coatings to prefinish fiberboard and particleboards. For. Prod. J. 12(12): 585–588.

———. 1973. Flat sheet hardboard finishing. For. Prod. J. 23(4): 26–29.

Johansen, T. A. 1971. Particleboard bonded with sulphite liquor. Fifth Pclebd. Sympo. Proc., Wash. St. Univ., Pullman: 11–30.

Johnson, E. S., ed. 1956. Wood particleboard handbook. School of Eng., N.C. St. Coll., Raleigh.

Johnson, J. W. 1956. Dimensional changes in hardboard from soaking and high humidity. For. Prod. Res. Lab., Oreg. St. Univ., Corvallis, Rept. No. T–16.

———. 1964. Effect of exposure cycles on stability of commercial particleboard. For. Prod. J. 14(7): 277–282.

Johnson, R. E. 1968. Plant experiences and problems with fire retardants. Second Pclebd. Sympo. Proc., Wash. St. Univ., Pullman: 123–127.

Johnson, W. P. 1964. Flame-retardant particleboard. For. Prod. J. 14(6): 273–276.

Jones, W. A. 1968. Principles, problems, and techniques for cutting particleboard. Second Pclebd. Sympo. Proc., Wash. St. Univ., Pullman: 143–151.

Jorgensen, R., Murphey, W. K. 1961. Particle geometry and resin spread: its effect on thickness and distribution of glue line in oak particleboard. For. Prod. J. 11(12): 582–585.

———, and Odell, R. L. 1961. Dimensional stability of oak flakeboard as affected by particle geometry and resin spread. For. Prod. J. 11(10): 463–366.

Kalina, H. 1972. Rheological behavior and fatigue strength of plywood, particle-boards, and hardboards. Holztechnol. 13(3): 172–175.

Kamrath, R. H. 1963. Performance of phenolic and urea resins in particleboard binding. Adhesion, ASTM Special Tech. Pub. No. 360: 115–122.

Karger, S., and Bohme, P. 1971. Properties of plastic laminated particleboards. Part 1. Holzindustrie 24, No. 1, pp. 12–16, and No. 2, pp. 37–38.

Keaton, C. M., and Berger, R. E. 1959. Developing the best resin for the particular process. For. Prod. J. 9(3): 29A–30A.

Keesey, L. H. 1970. Plant experiences producing medium-density fiberboard. Fourth Pclebd. Sympo. Proc., Wash. St. Univ. Pullman: 227–230.

Kehr, E. 1962. Investigation on the suitability of various wood species, and assortments for chipboard manufacture. Part 4: Alder. Holztechnol. 3(2): 130–136.

———. 1964. Investigation of the suitability of various types and categories of wood for the manufacture of particleboard. Parts 3 and 4. USDA Transl. FPL–603.

———. 1966. On the improvement of particleboard surfaces. Holz Roh-u. Werkstoff 24(7): 295–304.

———. 1972. Assessment of the quality of coarse chips and of chips and particleboards manufactured from coarse chips. Part 1: Processing coarse chips in the particleboard industry, and assessment of the quality of coarse chips. Part 2: Quality of the chips obtained from coarse chips and properties of the particleboards derived from coarse chips. Holzindustrie 25(3,4): 70–73, 101–105.

———, and Boehme, P. 1962. Resistance of particleboards in humid-warm climates. USDA Transl. FPL–591.

———, and Glumann, H. D. 1970. Variations in the thickness of particleboard after pressing. Holztechnol. 11(1): 25–31.

———, and Jensen, V. 1970. Manufacture and properties of particleboard with five particle surface layers. Holz Roh-u Werkstoff 28(10): 385–591.

———, and Scherfke, R. 1970. Deformation of three-layer particleboards. Holztechnol. 11(4): 258–263.

———, and Schilling, F. 1965. Investigations on the suitability of various wood species and assortments for chipboard manufacture. Part 6: Birch. Holztechnol. 6(3): 161–168.

———, and ———. 1965. Investigations on the suitability of various wood species and assortments for chipboard manufacture. Part 7: Oak, aspen, poplar, hornbeam, elm, larch (compared with spruce and Scotch pine). Holztechnol. 6(4): 225–232.

———, and Schoelzel, S. 1966. The investigation of pressing conditions in the manufacture of particleboard. USDC Transl. FPL–678.

———, and ———. 1968. Studies on the pressure diagram for the manufacture of particleboard. Part 2: Effect of chip moisture, mold closing time, and molding pressure on the sealing characteristics in the hot pressing of particleboard. USDA Transl. FPL–689.

———, Macht, K. H., and Riehl, G. 1965. On the chip bonding process in particleboard manufacture. USDA Transl. FPL–610.

Kelly, M. W. 1970. Formaldehyde odor and release in particleboard. Fourth Pclebd. Sympo. Proc., Wash. St. Univ., Pullman: 137–150.

Keylwerth, R. 1958. On the mechanics of the multi-layer particleboard. Holz Roh-u Werkstoff 16(11): 419–430.

Kimoto, R., Ishimori, E., Sasaki, H., and Maku, T. 1964. Studies on particleboards.

Part 6: Effects of resin content and particle dimension on the physical and mechanical properties of low-density particleboards. Wood Res., Kyoto Univ. 32: 1–14.

Kitahara, K., and Mizuno, Y. 1961. Relationship between tree species and properties of particleboard. J. Japan Wood Res. Soc. 7(12): 239–241.

Klauditz, W. 1952. Investigation of the suitability of different types of wood, especially copper beech for the manufacture of particleboard. Tech. Inst., Inst. Wood Res., Rept. 25, Braunschweig, W. Germany.

———. 1954. Determination of urea-formaldehyde resin content (adhesive) in wood particleboards. Tech. Inst., Inst. Wood Res., Rept. 40/45, Braunschweig, W. Germany.

———. 1962. On the development and the situation of particleboard manufacture from 1955 to 1961. Holz Roh-u Werkstoff 20(1): 1–12.

———, and Kratz, W. 1962. Investigations on the possibilities of manufacture and the properties of simple wooden particle moldings. Holz Roh-u Werkstoff 20(1): 39–48.

———, and Meier, K. 1960. Determination of the percentage of urea and melamine resins in wood particleboards. Holz Roh-u Werkstoff. 18(5): 163–166.

Kleinlogel, A. 1950. Influences of concrete. Frederick Ungar, New York.

Klinkmuller, H. 1972. Continuous measurement and control of moisture content in particleboard manufacture. Holz Roh-u Werkstoff 30(12): 547–563.

Klippert, R. V. 1971. Design, installation, and operation of bag houses for particleboard plants. Fifth Pclebd. Sympo. Proc., Wash. St. Univ., Pullman: 275–299.

Knapp, H. J. 1968. Development, principles, problems, and techniques of finishing particleboard. Second Pclebd. Sympo. Proc., Wash. St. Univ., Pullman: 201–256.

———. 1971. Continuous blending 1948–71. Fifth Pclebd. Sympo. Proc., Wash. St. Univ., Pullman: 45–56.

Kolb, H. 1971. Properties of hard fiberboards in relation to their use in building. Holz Roh-u Werkstoff 29(1): 10–23.

Kollman, F. 1954. The present state of technology of waste wood board manufacture. Holz Roh-u Werkstoff. 12(4): 117–134.

———. 1957. Resin application and equipment for blending resin with raw material particles. FAO/ECE/Board Cons. Paper 5–30.

———. 1966. Wood particleboards. Springer, Berlin, pp. 424–440.

———. 1972. Creeping of wood and wood-based materials. Holztechnol. 13(2): 88–95.

———, and Fengel, D. 1965. Changes in the chemical composition of wood by thermal treatment. Holz Roh-u Werkstoff 23(12): 461–468.

———, and Schneider, A. 1963. On the sorption behavior of heat stabilized wood. Holz Roh-u Werkstoff 21(3): 77–85.

Korcago, I. G., and Panjukova, G. A. 1970. The equilibrium moisture content of particleboards. Derev. Prom. 19(5): 19–20.

Kornblum, G. 1956. Particleboards from flax. Part 2: Industrial experience in the use of flax straw for the manufacture of particleboards. FAO/ECE/Board Cons. Paper 4–16.

Koukal, M. 1969. Preservative treatment of wood-based materials. Mater. u. Organ., Beih. No. 2: 81–90.

Kozel'cev, L. I. 1965. Pressing particleboard under RF heating. USDA Transl. FPL–615.

Kraemer, E., and Schmitz-Lettanne, A. 1971. European laminating practices. Fifth Pclebd. Sympo. Proc., Wash. St. Univ., Pullman: 71–98.

Krebs, H. 1967. Wood-chip moldings. Holz Roh-u Werkstoff 25(10): 383–392.

Kubiak, M., Balazy, S., and Dymalski, E. 1969. The mycoflora of Scotch pine chips stored in conical piles. Drev. Vyskum (1): 11–23.

———, ———, and ———. 1970. Some rare and little known species of fungi found in piles of Scotch pine chips stored in the open. Drev. Vyskum (1): 15–20.

Kubinsky, E., and Ifju, G. 1973. A simple and fast method of pH measurement for wood. For. Prod. J. 23(2): 54–56.

Kubler, H. 1969. Bowing of panels in one-sided atmospheres. For. Prod. J. 19(11): 43–49.

———. 1969. Use of particleboard in West Germany. For. Prod. J. 19(9): 29.

Kuhlmann, G. 1962. Investigations of the thermal properties of wood and particleboard in dependency from moisture content and temperature in the hygroscopic range. Holz Roh-u Werkstoff 20(7): 259–270.

Kulinicheva, I. A., and Petri, V. N. 1971. The warping of lignocellulosic wood-based sheets. Lesn. Z. 14(2): 77–81.

Kumar, A., Gupta, R. C., and Jain, N. C. 1970. Building boards from barks. Indian Pulp Pap. 25(1/6): 414–419.

———, ———, and ———. 1971. Fiberboards from *Shorea robusta* (Sal.) bark. Indian For. 97(7): 422–429.

Kumar, V. B. 1961. Commercial manufacture of hardboard from mixed hardwoods. Part 1. Norsk Skogsindustri 15(5): 239–244.

Kunesh, R. H. 1960. The inelastic behavior of yellow poplar in compression perpendicular to the grain. M.S. thesis, Univ. of Mich., Ann Arbor.

Kuster, W. V., and Mallory, J. B. 1962. Gypsum, minerals yearbook. U.S. Bureau of Mines, Wash., D.C.

Laidlow, R. A., and Hudson, P. W. 1969. Tour of chipboard factories and research institutes in Europe with special reference to the production of exterior quality boards. Min. of Tech., For. Prod. Res. Lab., Princes Risborough, England.

Lamberts, K., and Pungs, L. 1962. Application of radio frequency heat in particleboard manufacture. Holz Roh-u Werkstoff 20(10): 405–408.

Lambuth, A. L. 1967. Performance characteristics of particleboard resin binders. First Pclebd. Sympo. Proc., Wash. St. Univ., Pullman: 219–233.

Larmore, F. D. 1959. Influence of specific gravity and resin content on properties of particleboard. For. Prod. J. 9(4): 131–134.

Larson, M. L., and Yen, M. M. 1969. Meeting fire hazard requirements with wood-base panel boards. For. Prod. J. 19(2): 12–16.

Lawniczak, M., and Nowak, K. 1962. Effect of water repellent on the deformations in particleboards and flaxboard caused by moisture changes. Holz Roh-u Werkstoff. 20(2): 68–71.

Lea, G. M. 1956. The chemistry of cement and concrete. Edward Arnold, London.

Lebedev, V. S., Golubov, I. A., and Prokofev, N. S. 1971. Effect of technological factors on the acoustic and physical and mechanical properties of particleboards. Derev. Prom. 20(6): 12–15.

Lehman, W. F. 1964. Retarding dimensional changes in particleboards. For. Prod. Res. Lab., Oreg. St. Univ., Corvallis, Info. Cir. No. 20.

———. 1965. Improved particleboard through better resin efficiency. For. Prod. J. 15(4): 155–161.

———. 1968. Durability of exterior particleboard. Second Pclebd. Sympo. Proc., Wash. St. Univ., Pullman: 275–306.

———. 1968. Molding compounds from Douglas-fir bark. For. Prod. J. 18(12): 47–53.

———. 1968. Resin distribution in flakeboard shown by ultraviolet light photography. For. Prod. J. 18(10): 32–34.

———. 1970. Resin efficiency in particleboard as influenced by density, atomization and resin content. For. Prod. J. 20(11): 48–54.

———. 1972. Moisture-stability relationships in wood-base composition boards. For. Prod. J. 22(7): 53–59.

Lengel, D. L. 1967. The Reitz dryer. First Pclebd. Sympo. Proc., Wash. St. Univ., Pullman: 183–193.

Leppin, J., Lamberts, K., and Pungs, L. 1966. Recent investigations on shortening the pressing time in chipboard production. USDA Transl. FPL–688.

Lewis, G. D. 1969. Market outlook for particleboard and hardboard. For. Prod. J. 19(9): 49–51.

Lewis, W. C. 1956. Testing and evaluating procedures for building boards. For. Prod. J. 6(7): 241–246.

———. 1958. Use development for particleboard. For. Prod. J. 8(2): 27A–30A.

———. 1967. Insulation board, hardboard, and particleboard. USDA Forest Service, For. Prod. Lab. Rept. FPL–077.

———. 1967. Thermal conductivity of wood-base fiber and particle panel materials. USDA Forest Service, For. Prod. Lab. Rept. FPL–77.

———. 1971. Board materials from wood residues. USDA Forest Service, For. Prod. Lab. Res. Note FPL–045.

———, and Heebink, B. G. 1971. Reconstituted products from oak. Oak Sympo. Proc., USDA Forest Service, NEFES, Upper Darby, Pa.: 106–110.

Lichtenberg, W. 1959. Investigation of the consequences in regard to transverse tensile strength, bending strength, and thickness swelling resulting from utilizing pole wood as material for inner ply of an otherwise normal chipboard panel. Unpublished thesis, Inst. Wood and Fiber Tech., Inst. of Tech., Dresden, E. Germany.

Liiri, O. 1960. Investigation on the effect of moisture and wax upon the properties of wood particleboard. Paperi ja Puu. 42(2) : 43–45, 52–56.

———. 1961. Investigation of long-term humidification and the effect of alternate moisture variations on the properties of particleboards. Helsingfors Statena tekniska forskningsanstatt Tiedotus Series I Puu. No. 21: 1–31.

———. 1969. The most recent developments in the wood particleboard line. USDC Transl. FPL–693.

————. 1969. The pressure in the particleboard production. Holz Roh-u Werkstoff. 27(10): 371–378.

————. 1971. Pressure in particleboard manufacture. New Zealand Forest Service, Wellington. pp. 13.

Loewenfeld, V. K. 1964. Particleboard presses with tray infeed. Trans. G. H. Buckling. Borden, Inc., Adh. & Chem. Div., Ser. Bull. SB–109, Sept. 1966.

Lofgren, B. E. 1957. Continuous and discontinuous heat treatment and humidi- fication of hardboards. In Fiberboard and particleboard, Vol. 3. FAO, Rome.

Loock, G. E. 1970. Space age fire detection. Fourth Pclebd. Sympo. Proc., Wash. St. Univ., Pullman: 183–196.

Louis, R. L. 1956. Producing particleboard by a continuous process. For. Prod. J. 6(5): 176–179.

Loycke, H. J. 1961. The utilization of low-grade wood assortments. Holz-Zbl. 87(47): 705–706.

Lundgren, A. S. 1969. Wood-based panel products as building materials. Swedish Wallboard Manufacturers Assn., Swedish Fiberboard Information 2.11, Part 1.

Luthardt, H. 1972. Gluing of wood particles with phenolic resin powdered glue. Holztechnol. 13(3): 135–139.

Lutz, J. F., Heebink, B. G., Panger, H. R., Hefty, F. V., and Mergen, A. F. 1969. Surfacing softwood dimension lumber to produce good surfaces and high value flakes. For Prod. J. 19(2): 45–51.

Lynam, F. C. 1959. Factors affecting the properties of wood chipboard. J. Inst. Wood Sci. No. 4: 14–27.

————. 1959. Manufacture and use of particleboards. Board 2: 21–26.

McCarthy, E. T., and Parkinson, J. R. Measuring moisture for particleboard manufacture. Third Pclebd. Sympo. Proc., Wash. St. Univ., Pullman: 45–60.

MacDonald, M. D. 1951. Compression of Douglas-fir veneer during pressing. Timberman 52(4): 98–100; (5): 86, 88, 90, 92.

McGee, L. B., McLean, R. A., and Carlyle, A. A. 1957. Properties of particleboard related to its use in furniture manufacture. For. Prod. J. 7(3): 91–95.

MacLean, J. D. 1941. Thermal conductivity of wood. Heat, Piping & Air Cond. 13(6): 380–391.

McNamara, W. S., and Shaw, M. D. 1972. Vacuum-pressure impregnation of medium-density hardboard with phenolic resin. For. Prod. J. 22(11): 19–22.

McNeil, W. M. 1967. Defiberizors for particleboard. First Pclebd. Sympo. Proc., Wash. St. Univ., Pullman: 117–134.

Maku, T., and Hamada, R. 1955. Studies on chipboard. Part 2: Swelling properties of chipboard. Wood Res., Kyoto Univ. 15: 38–59.

————, ————, and Sasaki, H. 1957. Studies on the chipboard. Part 3: Some experiments on the improvement of dimensional stabilities of chipboard. Wood Res., Kyoto Univ., No. 17.

————, ————, and ————. 1959. Studies of particleboard. Part 4: Temperature and moisture distribution in particleboard during hot pressing. Part 5: Influence of press time on thickness and cleavage strength of particleboard. Wood Res., Kyoto Univ. 21: 34–50.

————, Sasaki, H., and Hamada, R. 1956. The effect of paraffin emulsion on the hygroscopic, swelling, and mechanical properties of chipboard. J. Japan Wood Res. Soc. 2(3): 130–132.

Maloney, T. M. 1969. Resin distribution in layered particleboard. Quest, Wash. St. Univ., Coll. of Eng., Pullman 7(3): 28.

————. 1970. Resin distribution in layered particleboard. For. Prod. J. 20(1): 43–52.

————. 1971. The new environment in particleboard plants. For. Prod. J. 22(6): 11–14.

Marian, J. E. 1956. A captive plant for production of synthetic resin adhesives and binders. Swedish For. Prod. Lab., Stockholm.

————. 1958. Adhesives and adhesion problems in particleboard production. For. Prod. J. 8(6): 172–176.

————. 1960. The utilization of barks. Part 4: Fibers from spruce bark in wet process hardboard. Svensk Papperstidn. 62(7): 225–229.

————. 1967. Surface texture. In Adhesion and adhesives, Vol. 2, ed. R. Houwink and G. Salomon. Elsevier, Amsterdam and New York.

————. 1967. Wood, reconstituted wood and glued laminated structures. In Adhesion and adhesives, Vol. 2, ed. R. Houwink and G. Salomon. Elsevier, Amsterdam and New York.

————, and Wissing, A. 1956–1957. The utilization of bark: Index to bark literature. Svensk Papperstidn. 59(21): 751–758, (22): 800–805; 60(2): 45–49, (3) 85–87, (4): 124–127, (5): 170–174, (7): 255–258, (9): 348–352, (11): 420–424, (14): 522–523.

Marra, G. G. 1954. Discussion of article by Turner. For. Prod. J. 4(10): 210–223.

————. 1956. Some innovations in methods of producing special particles for board manufacture in the United States. FAO/ECE/Board Cons. Paper 5–28.

————. 1958. Particleboards: their classification and composition. For. Prod. J. 8(12): 11A–16A.

————. 1967. Introduction. First Pclebd. Sympo. Proc., Wash. St. Univ., Pullman: 7–9.

————. 1968. Introduction. Second Pclebd. Sympo. Proc., Wash. St. Univ., Pullman: 3–5.

————. 1969. Introductory remarks: A vision of the future for particleboard. Third Pclebd. Sympo. Proc., Wash. St. Univ., Pullman: 3–8.

————. 1970. Small steps for particleboard, a giant stride for the wood industry. Fourth Pclebd. Sympo. Proc., Wash. St. Univ., Pullman: 281–285.

Martin, J. T. 1967. Surface treatment of adherents. In Adhesion and adhesives, Vol. 2, ed. R. Houwink and G. Salomon. Elsevier, Amsterdam and New York.

Martin, R. E. 1963. Thermal properties of bark. For. Prod. J. 13(10): 419–426.

————. 1967. Interim equilibrium moisture content values of bark. For. Prod. J. 17(4): 30–32.

————. 1969. Characteristics of southern pine barks. For. Prod. J. 19(8): 23–30.

Matsuda, T., and Sano, Y. 1971. Manufacture of dry processed hardboard using waste fibers rejected from pulp mill. Wood Ind. 26(9): 22–25.

Maxey, C. W. 1962. Instrumentation for the determination of surface texture and

contact area as variables in the gluing of wood. M. S. thesis, Univ. of Calif.,
Berkeley.

————. 1962. Measuring texture and contact area of surfaces in gluing wood.
M.S. thesis, Univ. of Calif., Berkeley.

Maxwell, J. W. 1956. Resin for the manufacture of particleboards. FAO/ECE/
Board Cons. Paper 4–19.

————. 1969. Automatic handling of chemicals in particleboard plants. Third
Pclebd. Sympo. Proc., Wash. St. Univ., Pullman: 73–89.

May, H. A. 1970. Effects of press control on efficiency and quality of particleboard
production. Holz Roh-u Werkstoff 28(10): 391–396.

Megson, J. J. L. 1958. Phenolic resin chemistry. Academic Press, New York.

Meinecke, E. 1960. Physical and technical events in applying glue to wood chips
during particleboard manufacture. Ph.D. diss., Tech. Inst., Inst. Wood Research, Braunschweig, W. Germany.

————. 1962. On the physical and technical processes for the gluing and adhesion of
wood flakes for the production of wood particleboard. Tech. Inst., Inst. Wood
Research, Rept. 1053, Braunschweig, W. Germany.

Mestdagh, M., and Demeulemeester, M. 1970. Production of phenolic-resin-bonded
flaxboards for the building industry. Holz Roh-u Werkstoff 28(6): 209–214.

Meyer, F. J., and Spalding, D. H. 1956. Anti-termite and anti-fungal treatment for
fiberboard and particleboard. FAO/ECE/Board Cons. Paper 5–24.

Midyette, A. L. 1957. Wood Particle molding. For. Prod. J. 7(1): 1–6.

Miels, G., and Scheibert, W. 1957. Shortening of compacting time. Holzindustrie
10(5): 162.

Miller, D. J. 1972. Molding characteristics of some mixtures of Douglas-fir bark
and phenolic resin. For. Prod. J. 22(9): 67–70.

Mohr, E. 1964. The metal changing plate in particleboard plant. Holz-Zbl. 90,
No. 46, Beilage Moderne Holzverarbeitung.

Morgan, P., ed. 1961. Glass reinforced plastics. 3rd ed. Iliffe, London.

Morschauser, C. 1969. How to specify particleboard. Woodworking Digest.
September issue.

Morse, E. H. 1967. Washington Iron Works presses. First Pclebd. Sympo. Proc.,
Wash. St. Univ., Pullman: 359–380.

Moskvitin, N. I. 1969. Physiochemical principles of gluing and adhesion processes. USDA–NSF Transl. USDA Clearing House for Sci. & Tech. Infor.,
Springfield, Va.

Mossberg, B. 1957. Oil tempering of hardboard. In Fiberboard and particleboard.
FAO, Rome.

Mottet, A. L. 1967. The Buttner jet dryer. First Pclebd. Sympo. Proc., Wash.
St. Univ., Pullman: 175–181.

————. 1967. The particle geometry factor in particleboard manufacturing. First
Pclebd. Sympo. Proc., Wash. St. Univ., Pullman: 23–73.

Muller, H. 1962. Experience with paraffin wax emulsions as antiswelling agents
in chipboard industry. Holz Roh-u Werkstoff 20(11): 434–437.

Munk, E. 1960. Potentialities of wood chip molding. Timber Technology 68(5):
182–184, 68(6): 214–216.

Murdock, R. G. 1968. Principles, problems, and techniques for sanding particle-board. Second Pclebd. Sympo. Proc., Wash. St. Univ., Pullman: 165–179.

———. 1970. Complete panel system. Fourth Pclebd. Sympo. Proc., Wash. St. Univ., Pullman: 123–136.

Murphy, W. G., and Rishel, L. E. 1969. Relative strengths of boards made from bark of several species. For. Prod. J. 19(1): 52.

Myers, G. C. 1971. What's in the wastepaper fiber collected from municipal trash. Paper Trade J. August issue.

Narayanamurti, D., and Bist, B. S. 1964. Building boards from bamboos. Composite Wood 1(2): 9–54.

———, Gupta, B. C., and Verma, G. M. 1962. Influence of extractives on the setting of adhesives. Holzforsch. u. Holzverwert. 14(5/6): 85–88.

Neusser, H. 1964. About the changes of glues during the production process of particleboard and some factors influencing their adhesive quality. USDA Transl. FPL–546.

———, and Krames, U. 1971. Surface texture of particleboards, its assessment and its aesthetic effect. Holz Roh-u Werkstoff 29(3): 103–118.

———, and Schall, W. 1970. Experiments to study hydrolysis phenomena in chipboards. Holzforsch. u. Holzverwert. 22(6): 116–120.

———, and ———. 1970. Resistance of some types of particleboard and fiber-board to fungi. Holzforsch. u. Holzverwert. 22(2): 24–40.

———, and Zentner, M. 1970. Weathering experiments with chipboards and fiberboards. Holzforsch u. Holzverwert. 22(3): 50–60.

———, and ———. 1971. Weathering experiments with particle and fiberboards. USDA Transl. FPL–716.

———, Krames, U., and Haidinger, K. 1965. The response of particleboard to moisture with special regard to swelling. USDC Transl. FPL–658.

——— et al. 1969. Influence of the chip on the surface quality of chipboard. USDA Transl. FPL–710.

Nevers, A. D., and Lehmann, W. F. 1969. Triethylamine stabilizer for resins in Douglas-fir particleboard. For. Prod. J. 19(2): 53–59.

Newman, W. M. 1967. Principles of mat formation. First Pclebd. Sympo. Proc., Wash. St. Univ., Pullman: 311–317.

———. 1967. The Bahre-Bison forming machine. First Pclebd. Sympo. Proc., Wash. St. Univ., Pullman: 333–338.

Nicholson, D. C. 1971. Characteristics of particulate emissions from particleboard processes. Fifth Pclebd. Sympo. Proc., Wash. St. Univ., Pullman: 195–208.

Nicht, J., Luthardt, H., and Gunther, B. 1971. Conversion of sawdust and bark into board products. Holzindustrie 24(7): 199–204.

———, ———, and ———. 1971. Conversion of sawdust and bark into board products. Holzindustrie 24(11): 326–329.

Noack, D., and Schwab, E. 1972. The shear strength of particleboard as a criterion of particle bonding. Holz Roh-u Werkstoff 30(11): 430–444.

Northcott, P. L. 1959. Case-hardening of plywood. For. Prod. J. 9(9): 442.

Oberlein, A. 1956. Particleboards as panel materials. FAO/ECE/Board Cons. Paper 7–8.

————. 1957. Particleboards as panel materials. FAO/ECE/Board Cons. Paper 7–8.

Odell, F. G. 1971. Air quality standards for particleboard plants. Fifth Pclebd. Sympo. Proc., Wash. St. Univ., Pullman: 141–169.

Ogland, N. J. 1957. The heat treatment of hardboard. In Fibreboard and particleboard, Vol. 3. FAO, Rome.

Oleesky, S. S., and Mohr, J. G. 1964. Handbook of reinforced plastics. Reinhold, New York.

Olson, D. E. 1960. Studies to improve the dimensional stability of particleboard. M.S. thesis, Chem. Eng. Dept., Univ. of Wis., Madison.

Organ, E. 1967. Siempelkamp closed-frame, multi-opening press for particleboard. First Pclebd. Sympo. Proc., Wash. St. Univ., Pullman: 381–389.

Orth, H., Nahrich, K. H., and Schall, H. 1970. Vibration behavior of polymer-particleboard. Holz-Zbl. 96(105): 1517–1518.

Ota, M., and Tsutsumi, J. 1959. Studies on the particleboard produced from planer shavings. Part 2: The influence of the curing cycle and the beginning time of the breathing period on the properties. J. Japan Wood Res. Soc. 5: 92–95.

Otlev, I. A. 1969. Guide to particleboards. Izdatelstvo Lesnaja Promyselnnost, Moscow. 168 pp.

————. 1971. Change in the moisture content of the chip mat during hot pressing. Derev. Prom. 20(10): 3–4.

————. 1972. A method of working out pressing schedules for particleboards. Derev. Prom. 21(5): 5–7.

Pagel, H. F. 1967. Molding wood particles. First Pclebd. Sympo. Proc., Wash. St. Univ., Pullman: 405–430.

Pahlitzsch, G., and Mehrdorf, J. 1962. Production of wood chips with disk cutters. Holz Roh-u Werkstoff. 20(8): 314–422; 20(9): 408–418; 20(11): 443–453; 21(4): 144–149 (1963).

————, and Sommer, I. 1966. Production of wood chips with a cylindrical cutter. USDA Transl. FPL–677.

Pana, G. I. 1970. Particleboards and fiberboards with characteristics appropriate to their uses. Industr. Lemn. 21(10): 361–367.

Papreckis, B. A., and Pesockij, A. N. 1971. Standardization of panel dimensions for the centralized sawing of particleboard. Derev. Prom. 20(1): 8–11.

Patent (U.S.) 1951. Patent No. 2,571,986.

Patent (U.S.) 1951. Patent No. 2,572,070.

Patent (U.S.) 1953. Patent No. 2,629,674.

Patent (U.S.) 1954. Patent No. 2,686,143.

Patent (U.S.) 1954. Patent No. 2,697,677.

Patent (U.S.) 1965. Patent No. 3,166,617.

Patent (U.S.) 1965. Patent No. 3,202,743.

Patent (U.S.) 1966. Patent No. 3,271,492.

Patent (U.S.) 1969. Patent No. 3,478,861.

Patent (W. Germany) 1948. Patent No. 841,055.

Patent (W. Germany) 1959. Patent No. 1,049,084.

Patent (W. Germany) 1959. Patent No. 1,066,344.

Patterson, T. J., and Snodgrass, J. D. 1959. Effect of formation variables on properties of wood particle moldings. For Prod. J. 9(10): 330–336.

Paulitsch, M. 1972. Investigations on the pH-value of aqueous extractions from urea resin-glued particleboard. Holz Roh-u Werkstoff 30(11): 437–439.

Pecina, H. 1963. Determination of the fraction of surface area covered with adhesives for wood particles. Holztechnol. 4(1): 68–70.

———. 1965. Demonstration and action of the paraffin wax for water repellent processing of chipboard. USDA Transl. FPL–633.

Peinecke, R. 1967. Pre-pressing and the caulless system. First Pclebd. Sympo. Proc., Wash. St. Univ., Pullman: 339–345.

Perry, W. D. 1969. Thickness gauges for particleboard. Third Pclebd. Sympo. Proc., Wash. St. Univ., Pullman: 125–136.

Peters, C., and Cumming, J. D. 1970. Measuring wood surface smoothness: a review. For. Prod. J. 20(12): 40–43.

———, and Mergen, A. 1971. Measuring wood surface smoothness: a proposed method. For. Prod. J. 21(7): 28–30.

Peters, T. E. 1968. The vacuum former. Second Pclebd. Sympo. Proc., Wash. St. Univ., Pullman: 375–384.

———. 1969. Quality control in the particleboard plant. Third Pclebd. Sympo. Proc., Wash. St. Univ., Pullman: 15–25.

Petri, V. et al. 1971. Plastics made from wood of hardwood species. Lesn. Prom. 7: 14–15.

Pinion, L. C. 1967. Evaluation of urea-formaldehyde and melamine resins in particleboard. For. Prod. J. 17(11): 27–28.

———, and Farmer, R. H. 1958. Wet combustion of organic materials. Chem. & Ind. Ig.: 919–920.

Plath, E. 1971. Particleboard mechanics. Holz Roh-u Werkstoff 29(10): 377–387.

———. 1963. Influence of density on the properties of woodbase materials. Holz Roh-u Werkstoff 21(3): 104–108.

Plath, L. 1967. Tests on formaldehyde liberation from particleboard. Part 2: The influence of pressing time and temperature on formaldehyde liberation. Holz Roh-u Werkstoff 25(2): 63–68.

———. 1967. Tests on formaldehyde liberation from particleboard. Part 3: Influence of hardener coumpounds on formaldehyde liberation. Holz Roh-u Werkstoff 25(5): 169–173.

———. 1967. Tests on formaldehyde liberation from particleboard. Part 4: Effect of moisture content in chip mat on formaldehyde liberation. Holz Roh-u Werkstoff 25(6): 231–238.

———. 1968. Tests on formaldehyde liberation from particleboard. Part 5: The influence of curing acceleration and aging period on formaldehyde liberation. Holz Roh-u Werkstoff 26(4): 125–128.

———. 1968. Tests on formaldehyde liberation from particleboard. Part 6: Test series on the formaldehyde liberation from industrial particleboard. Holz Roh-u Werkstoff 26(11): 409–413.

———. 1971. Requirements for particleboards for lamination with plastics. Holz Roh-u Werkstoff 29(10): 369–376.

Plomley, K. F. 1966. Tannin formaldehyde adhesives for wood. Part 2: Wattle tannin adhesives. Council of Sci. and Ind. Res., Australia, Div. For. Prod. Tech. Paper No. 39. 16 pp.

Poller, S., Patscheke, G., Bertz, T., and Neumann, G. 1972. Influence of wood flour-grain size on the properties of wood flour-filled melamine resin-molding materials. Holztechnol. 13(3): 148–151.

Porter, S. M. 1971. Gas recycling and dryer modifications to reduce smoke emissions. Fifth Pclebd. Sympo. Proc., Wash. St. Univ., Pullman: 225–234.

Post, P. W. 1958. The effect of particle geometry and resin content on bending strength of oak particleboard. For. Prod. J. 8(10): 317–322.

———. 1961. Relationship of flake size and resin content to mechanical and dimensional properties of flakeboard. For. Prod. J. 11(1): 34–37.

Potutkin, G. F. 1971. Use of polyethene film in the manufacture of particleboards. Lesn. Z. 14(5): 170–171.

———, and Dranishnikov, G. L. 1971. The relation to chemical changes in wood constituents to the properties of particleboards. Lesn. Z. 14(1): 102–104.

Powell, H. L. 1968. Sandpaper for particleboard. Second Pclebd. Sympo. Proc., Wash. St. Univ., Pullman: 181–187.

Powers, P. O. 1943. Synthetic resins and rubbers. Wiley, New York.

Pozdnyakov, A. A. 1971. The degree of loading on differently oriented chips in particleboard. Lesn Z. 14(3): 84–87.

Pungs, L., and Lamberts, K. 1957. The application of high frequency heating in the particleboard industry. FAO/ECE/Board Cons. Paper 5–36.

———, and ———. 1962. On the economy of hardening thick and light particleboard by radio frequency heating. Holz Roh-u Werkstoff. 20(1): 49–51.

Putnam, R. D. 1965. Quality control in the particleboard industry. Remarks presented at the annual meeting of For. Prod. Res. Soc.: New York City.

Rachwitz, G. 1967. Sifting of wood particles. Part 3: Sifting in a variation air flow. Holz Roh-u Werkstoff. 25(5): 188–193.

———, and Obermaier, M. 1962. Sifting of wood particles. Part 1: Foundations of sifting and sifting in a horizontal air flow. Holz Roh-u Werkstoff. 20(1): 27–38.

Raddin, H. A. 1967. High frequency pressing and medium-density board. First Pclebd. Sympo. Proc., Wash. St. Univ., Pullman: 391–401.

———. 1970. The economics of the system producing dry process medium density fiberboard. Fourth Pclebd. Sympo. Proc., Wash. St. Univ., Pullman: 219–226.

Rauch, A. H., and Cheney, R. A. 1970. Pressurized refining of a variety of residuals. Fourth Pclebd. Sympo. Proc., Wash. St. Univ., Pullman: 237–252.

Rayner, C. A. A. 1967. Adhesive bonding processes. In Adhesion and adhesives, Vol. 2, ed. R. Houwink and G. Salomon. Elsevier, Amsterdam and New York.

———. 1967. Synthetic organic adhesives. In Adhesion and adhesives, Vol. 1, ed. R. Houwink and G. Salomon. Elsevier, Amsterdam and New York.

Reason, R. E. 1960. The measurement of surface texture. Cleaver-Hume Press.

Rice, J. T. 1973. Particleboard from "silage" sycamore-laboratory production and testing. For. Prod. J. 23(2): 28–34.

Riedell, A. W. 1971. Factors in converting to super finishes. For. Prod. J. 21(5): 12–14.

Riehl, G., Rajkovic, E., Kubin, J., and Kehr, E. 1972. Technological tests with various phenolic resins for the manufacture of particleboards and a comparative evaluation of their reactivity. Holztechnol. 12(2): 75–80.

——, ——, ——, and ——. 1972. Technological tests with various phenolic resins for the manufacture of particleboards and a comparative evaluation of their reactivity. Holztechnol. 13(1): 14–19.

Roffael, E., and Rauch, W. 1971. Production of particleboard with sulphite black liquor as binder. Part 1: Literature review and present investigations. Holzforschung 25(4): 112–116.

——, and ——. 1971. Production of particleboards with sulphite black liquor as binder. Part 2: A new and rapid method for production of particleboards bonded with sulphite waste liquor. Holzforschung 25(5): 149–155.

——, and ——. 1972. Influence of density on the swelling behavior of phenolic resin bonded particleboards. Holz Roh-u Werkstoff 30(5): 178–181.

Rooney, J. E. 1970. Combining species. Fourth Pclebd. Sympo. Proc., Wash. St. Univ., Pullman: 231–236.

Ross, V. R. 1958. A statistical quality control program. For. Prod. J. 8(8): 24A–26A.

Roth, L. 1960. Structure, extractives, and utilization of bark. Inst. Paper Chem., Appleton, Wis., Bibliog. Series No. 191.

——. 1968. Structure, extractives, and utilization of bark. Inst. Paper Chem., Appleton, Wis., Suppl. No. 1.

——, and Wiener, J. 1961. Waxes, waxing, and wax modifiers. Inst. Paper Chem., Appleton, Wis., Bibliog. Series No. 198.

——, and ——. 1967. Barkers and barking of pulpwood. Inst. Paper Chem., Appleton, Wis., Suppl. No. 1.

——, Saeger, G., Lynch, F. J., and Weiner, J. 1960. Barkers and barking of pulpwood. Inst. Paper Chem., Appleton, Wis., Bibliog. Series No. 190.

Runkel, R. O. H., and Wilke, K. D. 1951. Contribution to knowledge about the thermoplastic behavior of wood. Holz Roh-u Werkstoff 9(7): 260–270.

Ryan, D. M. 1967. Equipment and methods for separating wood particles. First Pclebd. Sympo. Proc., Wash. St. Univ., Pullman: 135–159.

Saito, F. 1972. Springback of particleboards. Wood Ind. 27(1): 14–18.

Salomon, G. 1967. Adhesion. In Adhesion and adhesives, Vol. 1, ed. R. Houwink and G. Salomon. Elsevier, Amsterdam and New York.

——. 1967. Introduction. In Adhesion and adhesives, Vol. 2, ed. R. Houwink and G. Salomon. Elsevier, Amsterdam and New York.

Sandermann, W., and Augustin, H. 1964. Chemical investigations on the thermal decomposition of wood. Part 3: Chemical investigation on the course of decomposition. Holz Roh-u Werkstoff 22(10): 377–385.

Sandweg, K. 1957. 20 years of high frequency wood gluing. Holz Roh-u Werkstoff 15(4): 174–189.

Sarvis, J. C. 1959. Western bark utilization study. Western Pine Assn., Portland, Oreg., Res. Note 7–221.

Sattler, H. 1971. Possible applications and properties of building slabs made from cement and barking wastes. Holzindustrie 24(3): 68–70.

Saums, W. A., and N. Turner, H. D. 1961. Accelerated test for measuring lateral dimensional change of woodbase panels. For. Prod. J. 11(9): 406–408.

Scharf, G. 1972. Chain conveyers in particleboard manufacture, their use and construction. Holz Roh-u Werkstoff 30(2): 45–51.

Scharper, H. 1972. On the choice of heating platen thickness in particleboard presses. Holz Roh-u Werkstoff. 30(4): 127–129.

Schmidt, D. E. 1957. Some practical aspects of particleboard core stock. For. Prod. J. 7(10): 53A–54A.

Schmidt, H., and Nehm, A. B. 1972. Testing of particleboard on termite resistance. Holz Roh-u Werkstoff 30(5): 175–177.

Schmidt, L. 1958. Selection of mineralizer and the method of mineralization in making cement fibrolite. Stroitel Materialy 4: 12, 20–22.

Schmidt, W. 1972. Protection against fire, explosions, and pollution in the manufacture of particleboards. Holz-Zbl. 98(45): 5–12.

Schmidt-Hellerau, C. 1971. The suitability of some South American species for the manufacture of particleboards. Holz-Zbl. 97(11): 142–143.

Schmierer, R. E. 1971. Australian particleboard practices. Fifth Pclebd. Sympo. Proc., Wash. St. Univ., Pullman: 99–132.

Schmutzler, W. 1957. The utility of drum sanders on sanding of wood particleboard. Holz Roh-u Werkstoff 15(4): 170–174.

———. 1964. Feed systems on chippers. Holz Roh-u Werkstoff 22(6): 237–241.

———. 1966. Sanding machines for particleboard. Holz Roh-u Werkstoff 24(9): 390–395.

Schnitzler, E. 1966. Particle preparation and drying in manufacturing of particleboards. FAO/ECE/Board Cons. Paper 5–29.

———. 1971. New methods of chip gluing. Holz Roh-u Werkstoff 29(10): 382–389.

Schoring, P., and Stegmann, G. 1972. New wood-to-wood bonds in chip materials by chemical polyfunctional systems. Part 2: Hot pressing with wooden boards and particles in a weak acid bond system. Holz Roh-u Werkstoff 30(9): 329–332.

———, Roffael, E., and Stegmann, G. 1972. New wood-to-wood bonds in chip materials by chemical polyfunctional systems. Part 1: Hot pressing with amine-alkaline bond systems. Holz Roh-u Werkstoff 30(7): 253.

Schulman, J., and Winler, B. L. 1960. New uses for wood flour in rigid polyurethane foams. For. Prod. J. 10(10): 487–490.

Schumann, E. W. 1970. B. Maier-Black Clawson MKZ Flaker. Fourth Pclebd. Sympo. Proc., Wash. St. Univ., Pullman: 151–170.

Schwab, E. 1971. Particleboards containing bark. Forstarchiv. 42(10): 198–200.

Schwartz, S. L., and Baird, P. K. 1950. Effect of molding temperature on the strength and dimensional stability of hardboards from fiberized water-soaked Douglas-fir chips. Proc. For. Prod. Res. Soc. 4: 322–326.

Schwarz, F. E. 1971. Advances in detecting formaldehyde release. Fifth Pclebd. Sympo. Proc., Wash. St. Univ., Pullman: 31–44.

Sciconolfi, C. A. 1960. Resin solids in a saturated sheet. Tappi 43: 152A–153A.

Seborg, R. M., and Stamm, A. J. 1941. The compression of wood. USDA Forest Service, For. Prod. Lab. Rept. No. R1258.

———, Millett, M. A., and Stamm, A. J. 1956. Heat-stabilized compressed wood (Staypack). USDA Forest Service, For. Prod. Lab. Rept. No. 1580.

————, Tarkow, H., and Stamm, A. J. 1953. Effect of heat upon the dimensional stabilization of wood. J. For. Prod. Res. Soc. 3(3): 59–67.

Seifert, E., and Schmid, J. 1970. Temperature behavior of wood and wood-based materials in relation to coloring. Holz Roh-u Werkstoff 28(5): 178–182.

Seifert, K. 1959. The analysis of wood particleboards. Holz Roh-u Werkstoff 14(9): 328–332.

Seip, D. R. 1958. Economic aspects of a particleboard operation. For. Prod. J. 8(3): 26A–28A.

Selvidge, L. R. Jr. 1963. An experimental approach to compression molding wood particles. M.S. thesis, South. Ill. Univ., Carbondale.

Semana, J. A., and Anderson, A. B. 1968. Hardboards from Benquet pine bark-wood compositions. For. Prod. J. 18(7): 28–32.

Shen, K. C. 1970. Correlation between internal bond and the shear strength measured by twisting thin plates of particleboard. For. Prod. J. 20(11): 16–20.

————. 1970. Correlation between torsion-shear strength modulus of rupture of particleboard. For. Prod. J. 20(12): 32–36.

————. 1971. Evaluating particleboard properties by measuring saw-cutting force. For. Prod. J. 21(10): 46–52.

————. 1971. The relationship between torsion shear and other strength properties of particleboard. For. Prod. J. 21(6): 40–43.

————, and Carroll, M. N. 1970. Measurement of layer-strength distribution in particleboard. For. Prod. J. 20(6): 53–56.

————, and Fung, D. P. C. 1972. A new method for making particleboard fire-retardant. For. Prod. J. 22(8): 46–52.

————, and Wrangham, B. 1971. A rapid aging test procedure for phenolic particle-board. For. Prod. J. 21(5): 30–33.

Shumate, R. D. 1971. Design and operation of dry centrifugal dust collectors. Fifth Pclebd. Sympo. Proc., Wash. St. Univ., Pullman: 243–262.

Shvartsman, G. M., Svitkin, M. Z., and Zavya-ova, Z. V. 1971. New standards for wood particleboards. Derev. Prom. 20(6): 8–9.

Sieminski, R. 1966. Testing and measuring the surface structure of wood and wood-base materials. Holz Roh-u Werkstoff 24(9): 396–404.

Smart, D. W., and Cameron, R. E. 1971. Resistance of particleboard to *Poria monticola* and *Lenzites trabea*. New Zealand J. For. Sci. 1(2): 238–239.

Soine, H. 1972. Conveying and storing of raw material in the particleboard industry. Holz Roh-u Werkstoff 30(11): 414–421.

Soule, E. L., and Hendrickson, H. E. 1966. Bark fiber as a reinforcing agent for plastics. For. Prod. J. 16(8): 17–22.

Spalt, H. A., and Sutton, R. F. 1968. Buckling of thin surfacing materials due to restrained hygroexpansion. For. Prod. J. 18(4): 53–57.

Splawa-Neyman, S. 1970. Natural resistance of dry-felted fiberboards to fungi. Prace Inst. Tech., Drewna 17(3): 127–139.

————, and Szymankiewicz, H. 1970. Protection of dry-felted fiberboards made from Scotch pine wood against fungi and insects. Prace Inst. Tech., Drewna 17(3): 141–154.

Srivastava, L. M. 1964. Anatomy, chemistry, and physiology of bark. In International review of for. research, Vol. 1, ed. J. A. Romberger and P. Mikola. Academic Press, New York.

Stamm, A. J. 1959. Dimensional stabilization of wood by thermal reactions and formaldehyde cross-linking. Tappi 42(1): 39–44.

———. 1959. Effect of polyethylene glycol on the dimensional stability of woods. For. Prod. J. 9(10): 375–381.

———. 1961. A comparison of three methods for determining the pH of wood and paper. For. Prod. J. 11(7): 310–312.

———. 1962. Stabilization of wood; a review of current methods. For. Prod. J. 12(4): 158–160.

———. 1962. Wood and cellulose-liquid relationships. North Carolina Ag. Exp. Station, Tech. Bull. No. 150, pp. 21–24.

———, and Baechler, R. H. 1960. Decay resistance and dimensional stability of five modified woods. For. Prod. J. 10(1): 22–26.

———, and Cohen, W. E. 1956. Swelling and dimensional control of paper. Part 2: Effect of cyanoethylation, acetylation and cross-linking with formaldehyde. Australian Pulp. Pap. Ind. Tech. Assn. Proc. 10: 366–393.

———, Burr, H. K., and Kline, A. A. 1955. Heat stabilized wood (staybwood). USDA Forest Service, For. Prod. Lab. Rept. No. 1621.

Starzyska, K. 1964. Urea binders with low content of free formaldehyde, their properties and application. Holztechnol., Sonolerhoft Klebtechnik, S. 86.

Stayton, C. L., Hyvarinen, M. J., Gertjejansen, R. O., and Haygreen, J. G. 1971. Aspen and paper birch mixtures as raw material for particleboard. For. Prod. J. 21(12): 29–30.

Steck, E. F. 1970. Caulless pressing systems for the manufacture of particleboard. Fourth Pclebd. Sympo. Proc., Wash. St. Univ., Pullman: 99–122.

Stegmann, G., and Bismark, C. 1967. Manufacture of particleboards from hardwoods. Holz-Zbl. 93(76): 1211–1212; 93(78): 1213–1246; 93(81): 1313–1314.

———, and Durst, J. 1964. Beech pressboard. USDC Transl. FPL–638.

———, and Ginzel, W. 1965. Determination of the content of urea-formaldehyde adhesives in particleboard. USDC Transl. FPL–629.

———, and Kratz, W. 1967. Characterization of the bonding quality of particleboards of varying binder contents and densities through swelling pressure measurements. Part 1: The characterization of swelling tensions using a new measuring method. Adhesion 11(1): 11–18.

———, and May, H. A. 1968. Reducing the pressing time when making thick particleboards. USDA Transl. FPL–691.

Steiner, K. 1972. Chippers for long logs in the particleboard industry. Development and stand of technique. Holz Roh-u Werkstoff 30(6): 201–214.

Stensrud, R. K. 1968. Overlays for exterior particleboard. Second Pclebd. Sympo. Proc., Wash. St. Univ., Pullman: 307–341.

———, and Nelson, J. W. 1965. The importance of overlays to the forest products industry. For. Prod. J. 15(5): 203–205.

Stephenson, J. A. 1968. Fillers for particleboard. Second Pclebd. Sympo. Proc., Wash. St. Univ., Pullman: 257–268.

Stewart, D. L., and Butler, D. L. 1968. Hardboard from cedar bark. For. Prod. J. 18(12):19–23.

Stillinger, J. R. 1967. Drying principles and problems. First Pclebd. Sympo. Proc., Wash. St. Univ., Pullman: 163–173.

————. 1967. The Heil dryer. First Pclebd. Sympo. Proc., Wash. St. Univ., Pullman: 205–215.

Stofko, J. 1970. Particleboards with oriented particles. Drvna Ind. 21(6): 104–107.

Stojcev, A., and Novotny, M. 1966. Treatment of particleboard with molten paraffin wax to reduce hygroscopicity. Holztechnol. 7(2): 88–92.

Stolley, L. 1956. Some methods of treating particleboards to increase their resistance against fungi and termites. FAO/ECE/Board Cons. Paper 5–38.

Strickler, M. D. 1959. Effect of press cycles and moisture content on properties of Douglas-fir flakeboard. For. Prod. J. 9(7): 203–305.

Strisch, H., and Van Huellen, S. 1969. New systems and recent European developments. Third Pclebd. Sympo. Proc., Wash. St. Univ., Pullman: 229–250.

Sturos, J. A. 1973. Segregation of foliage from chipped tree tops and limbs. USDA Forest Service, For. Prod. Lab., Res. Note NC–146. 4 pp.

Suchsland, O. 1959. The strength of glue joints in wood obtained with minimum glue spread. Quart. Bull., Mich. Ag. Exp. Sta., Mich. St. Univ. 41(3): 577–599.

————, 1959. An analysis of the particleboard process. Quart. Bull. Mich. Ag. Exp. Sta., Mich. St. Univ. 42(2): 350–372.

————. 1962. The density distribution in flakeboards. Quart. Bull., Mich. Ag. Exp. Sta., Mich. St. Univ. 45(11): 104–121.

————. 1965. Swelling stresses and swelling deformations in hardboard. Quart. Bull., Mich. Ag. Exp. Sta., Mich. St. Univ. 47(4): 591–605.

————. 1966. Some performance characteristics of "interior" and "exterior" type particleboard. Quart. Bull., Mich. Ag. Exp. Sta., Mich. St. Univ. 49(2): 200–210.

————. 1967. Behavior of a particleboard mat during the press cycle. For. Prod. J. 17(2): 51–57.

————. 1971. Linear expansion of veneered furniture panels. For. Prod. J. 21(9): 90–96.

————. 1972. Linear hygroscopic expansion of selected commercial particleboards. For. Prod. J. 22(11): 28–32.

————, and Enlow, R. C. 1968. Heat treatment of exterior particleboard. For. Prod. J. 18(8): 24–28.

Sulzberger, P. H. 1943. The effect of temperature on the strength of wood, plywood, and glued joints. J. Council Sci. and Ind. Res., Australia 16(4): 263–265.

Surdyk, L. V. 1971. Dryer modifications to reduce smoke emissions and improve drying. Fifth Pclebd. Sympo. Proc., Wash. St. Univ., Pullman: 217–224.

Svarcman, G. M. 1970. The thermal properties of particleboards. Derev. Prom. 19(7): 8–9.

Syska, A. D. 1969. Exploratory investigation of fire-retardant treatments for particleboard. USDA Forest Service, For. Prod. Lab. Res. Note FPL–0201.

Szendrey, I. 1969. Role of wood chemical factors in the pressing of particleboard. Faipar, Budapest 19(4): 97–100.

Szymankiewicz, H., and Nowar, A. 1970. Examination of the pressing process in the manufacture of fiberboards by the dry process. Holztechnol. 11(1): 3–8.

Takahashi, H., Endo, K., and Suzuki, H. 1972. Studies on handling process of wood fiber for dry processing of fiberboard. Part 1: On particle size of ground fiber. Part 2: Particle size distribution of ground wood fiber. J. Japan Wood Res. Soc. 18(1): 9–19.

————, ————, and ————. 1972. Studies on handling process of wood fiber for dry processing of fiberboard. Part 3: Effect of grinding power required on size of fiber. J. Japan Wood Res. Soc. 18(2): 57–61.

Talbott, J. W. 1959. Flapreg flakeboard—resin impregnated compressed flakes. For. Prod. J. 9(2): 103–106.

————, and Maloney, T. M. 1957. Effect of several production variables on the modulus of rupture and internal bond strength of boards made of green Douglas-fir planer shavings. For. Prod. J. 7(10): 395–398.

Tarkow, H., and Stamm, A. J. 1953. Effect of formaldehyde treatments upon the dimensional stabilization of wood. J. For. Prod. Res. Soc. 3(2): 33–37.

Taylor, H. F. W., ed. 1964. The chemistry of cements. Academic Press, New York.

Teesdale, L. V. 1958. Thermal insulation made of wood-base materials—its application and use in houses. USDA Forest Service, For. Prod. Lab. Rept. 1740.

Temkina, R. Z. et al. 1971. Rapid-hardening binders for particleboards. Derev. Prom. 20(3): 3–5.

Textor, C. K. 1948. Wood fiber production with revolving disk mills. Proc. For. Prod. Res. Soc. 2.

Theien, C. M. 1970. Routing and shaping of particleboard. For. Prod. J. 20(6): 30–32.

Thomas, R. J., and Taylor, F. W. 1962. Urea-formaldehyde resins modified with water soluble blood. For. Prod. J. 12(3): 111–115.

Thompson, T. A. 1967. The Westvaco blender. First Pclebd. Sympo. Proc., Wash. St. Univ., Pullman: 291–293.

Timms, C. 1944. Measurement of surface waviness. Conf. on Surface Finish, Natl. Phys. Lab., Teddington.

Tomas, M. 1964. On the problematics of the release of formaldehyde upon hardening of urea-formaldehyde resins at increased temperature. Holztechnol. 4(5): 89.

Tomek, A. 1966. Heat treatment of wood chips; a new process for making particleboard water repellent. Holztechnol. 7(3): 157–160.

Traux, J. R. 1929. The gluing of wood. USDA Bull. No. 1500.

Treiber, H. 1965. Spatial adaptation of the gluing of high polymeric materials, especially wood particle mixtures. Holz Roh-u Werkstoff 22(8): 319–331.

Troger, J. 1971. A method for the objective evaluation of the edge quality of machined particleboards and fiberboards surface-treated by hot-pressing and plastic coating. Holztechnol. 12(2): 76–79.

Trotter, T. 1968. Sawblades for particleboard. Second Pclebd. Sympo. Proc., Wash. St. Univ., Pullman: 153–163.

Troughton, G. E. 1969. Effect of degree of cure on the acid-hydrolysis rates of formaldehyde glue-wood samples. J. Inst. Wood Sci. 23(7): 51–56.

Trutter, G., and Himmelheber, M. 1970. Modern particleboard manufacture: state of the art in various manufacturing processes. Transl. FPL–712.

Turner, H. D. 1954. Effect of size and shape on strength and dimensional stability of resin-bonded wood-particle panels. For. Prod. J. 4(5): 210–222.

Ulbricht, H. J. 1956. Equipment for manufacturing particles. FAO/ECE/Board Cons. Paper 5–27.

Urbanik, E. 1969. Investigations on the preservation of particleboard and fiberboard against fungi and insects. Mater. u. Organ., Beih No. 2: 95–102.

Vajda, P. 1967. Blenders, an introduction. First Pclebd. Sympo. Proc., Wash. St. Univ., Pullman: 269–273.

———. 1970. The economics of particleboard manufacture revisited or an assessment of the industry in 1970. Fourth Pclebd. Sympo. Proc., Wash. St. Univ., Pullman: 5–46.

Vakhrusheva, I. A., and Petri, V. N. 1964. Use of comminuted larch wood for making plastics without binders. USDA Transl. FPL–600.

Vanderbilt, B. M. 1964. The bonding of fiberous glass to elastomers. Rubber World. 150(4): 89.

Vodoz, J. 1957. The behavior of wood during drying by radio frequency heating. Holz Roh-u Werkstoff 15(8): 327–339.

Vorreiter, L. 1953. Basic properties of saw chips. Holz-Zbl. 79(105): 1127.

———. 1962. Manual of wood technology. Vol. 3. Vienna and Munich.

———. 1965. Particle physics as basis for the processing of (minute) wood chips. USDA Transl. FPL–605.

Wake, W. C. 1967. Rubbers. In Adhesion and adhesives, Vol. 1, ed. R. Houwink and G. Salomon. Elsevier, Amsterdam and New York.

Walter, F. and Knitsch, H. W. 1970. Contribution to testing the strength of edges of particleboard. Holztechnol. 11(1): 32–36.

Wangaard, F. F. 1950. The mechanical properties of wood. Wiley, New York.

Ward, R. J., and Skaar, C. 1963. Specific heat and conductivity of particleboard as functions of temperature. For. Prod. J. 13(1): 31–38.

Watson, D. A. 1959. Granulated wood molding. For. Prod. J. 9(8): 7A–8A.

———. 1959. Low-cost wood-particle moldings. Materials in Design Engineering 49(5): 103–104.

Watson, H. 1905. Composite board. U.S. Patent No. 796, 545.

Weatherwax, R. C., and Tarkow, H. 1964. Effect of wood on setting of portland cement. 14(12): 567–570.

———, and ———. 1967. Effect of wood on the setting of portland cement: Decayed wood as an inhibitor. For. Prod. J. 17(7): 30–32.

Weidas, Jr. N. C. 1965. The feasibility of particleboard manufacture in Bershire County, Massachusetts. Univ. of Mass. Bull., Amherst.

Wentworth, I. 1967. New designs for single opening presses. First Pclebd. Sympo. Proc., Wash. St. Univ., Pullman: 251–267.

———. 1968. Caulless process for making particleboard. For. Prod. J. 18(1): 12–13.

———. 1969. Molded products and laminates. Third Pclebd. Sympo. Proc., Wash. St. Univ., Pullman: 217–228.

———. 1969. New designs for single-opening presses. Third Pclebd. Sympo. Proc., Wash. St. Univ., Pullman: 251–267.

———. 1971. Production of particleboard in a continuous ribbon. Fifth Pclebd. Sympo. Proc., Wash. St. Univ., Pullman: 59–70.

West Coast Adhesive Manufacturers Association. 1966. A proposed new test for accelerated aging of phenolic resin-bonded particleboard. For. Prod. J. 16(6): 19–23.

————. 1970. Accelerated aging of phenolic resin-bonded particleboard. For. Prod. J. 20(10): 26–27.

White, J. T. 1970. The impact of plastics on the markets of wood in building and construction. For. Prod. J. 20(7): 12–15.

White, M. G. 1969. The resistance of chipboard to attack by wood boring insects. Mater. u. Organ., Beih. No. 2: 43–48.

————. 1970. Further experiments on the resistance of chipboard to attack by the common furniture beetle *Anobium punctatum*. B.W.P.A. News Sheet, Brit. Wood Pres. Assn. 112: 7.

Wiecke, P. H. 1970. Introduction to fire and explosion prevention and detection. Fourth Pclebd. Sympo. Proc., Wash. St. Univ., Pullman: 173–182.

Wilhelmi, H. 1957. The use of particleboards in building and construction work. FAO/ECE/Board Cons. Paper 7–9.

Willeitner, H. 1965. The behavior of wood particleboards under attack of basidio-mycetes. Part 1: Decomposition of particleboards by basidiomycetes. Holz Roh-u Werkstoff. 23(7): 264–270.

————. 1969. Laboratory testing of wood chipboards against fungal attack. Mater. u. Organ., Beih. No. 2: 109–122.

Witt, J. C. 1966. Portland cement technology. Second ed. Chemical Publishing Co., New York.

Wittmann, O. 1962. The subsequent dissociation of formaldehyde from particleboard. Holz Roh-u Werkstoff. 20(6): 221–224.

————. 1971. The manufacture of non-hygroscopic particleboard. New Zealand Forest Service, Wellington. pp. 14.

————. 1971. The manufacture of water-repellent particleboard. Holz Roh-u. Werkstoff 29(7): 259–264.

Wnuk, M. 1965. Controlled pressure variations in chipboard samples compared to pine sapwood samples. USDA Transl. FPL–611.

Wood, W. A. 1968. Cement-coated particles used in Japanese board. World Wood. August issue.

Worthington, A. G., and Halliday, D. 1948. Heat. Wiley, New York.

Yang, C. F., and Haygreen, J. G. 1971. Predicting flexural creep in particleboard. Wood & Fiber 3(3): 146–152.

Yashiro, M., Kawamura, Y., Sasaki, K., and Mamada, S. 1968. Studies on manu-facturing condition of wood wool-cement board. Part 1: Trial manufacturing of wood wool-cement board with unsuitable species. Mokuzai Kogyo 23(9): 19–22.

Yashiro, M. et al. 1970. Studies on conditions for the manufacture of wood wool-cement board. Part 1: Production of wood wool-cement board from unsuitable woods. USDA Transl. FPL–699.

Young, D. R. 1967. Engineering for general particleboard plant. For. Prod. J. 17(7): 10–12.

————. 1967. Remarks at the 1966 annual meeting of FPRS. For. Prod. J. 17(1): 66.

————. 1967. The Loedige blender. First Pclebd. Sympo. Proc., Wash. St. Univ., Pullman: 295–300.

————. 1967. The Ponndorf dryer. First Pclebd. Sympo. Proc., Wash. St. Univ., Pullman: 195–203.

Youngs, R. L. 1957. The perpendicular-to-grain mechanical properties of red oak as related to temperature, moisture content, and time. USDA Forest Service, For. Prod. Lab. Rept. No. 2079.

Yukna, A. D., Bankiyeris, Ya., and Ziyedin'sh, I. O. 1965. The flow of particle-glue mixture in pressing profiled parts from pulverized wood. USDA Transl. FPL–614.

Zhukov, V. P., and Mikailov, N. A. 1970. The strength of particleboards during conditioning. Derev. Prom. 19(6): 2–4.

————, ————, and Ostapenki, N. I. 1971. Cooling of particleboards after removal from the press. Derev. Prom. 20(7): 3–4.

Zukowski, L. 1961. Hydrophobating admixtures used in particleboard gluing. Przemysl. Drzewny, Warsaw 12(14): 11–12.

Index